PRINT CULTURE AND INTELLECTUAL LIFE IN IRELAND, 1660–1941

PRINT CULTURE AND INTELLECTUAL LIFE IN IRELAND 1660–1941

Essays in honour of Michael Adams

EDITED BY
Martin Fanning & Raymond Gillespie

THE WOODFIELD PRESS

This book was designed and typeset
in 11 on 14.5 Garamond by Martin Fanning
for *The Woodfield Press*
17 Jamestown Square, Dublin 8
www.woodfield-press.com
e-mail: terri.mcdonnell@ireland.com

ISBN 10-digit 1-905094-00-0
ISBN 13-digit 978-1-905094-00-4

A catalogue record for this title
is available from the British Library.

SPECIAL ACKNOWLEDGMENT
The publisher is very grateful to MPG Books
for generous sponsorship of this book.

Printed by
MPG Books Ltd, Cornwall.

Contents

Preface

HISTORIANS ARE WELL ACCUSTOMED to finding revolutions in nineteenth-century Irish society – revolutions in government, in devotion and in the practice of politics are all established ideas in the analytical armoury of those who have examined that world. Yet there is one other, and probably more fundamental, revolution that has remained uncharted. Ireland in the nineteenth century was engulfed in a wave of print that transformed almost every aspect of life. Almost every indicator of the activity of the world of print showed a marked expansion in the years after 1800. The number of printers and booksellers in Dublin increased significantly. Spurred by technical innovations in printing, large-scale publishing operations, such as the Dublin University Press, became more common than hitherto and outside Dublin, provincial printing also expanded. The newspaper business, for instance, showed a considerable growth: in 1850 there were sixty-five provincial newspapers which by 1892 had grown to 128. Evidence on circulation is more difficult to come by but it seems clear that increasing literacy, growing politicization and falling newspaper prices placed the output of this sector in more hands than ever before. This, however, is only part of the story. A rising number of books, published in England and elsewhere, also flooded into Ireland. The expansion of accessions to the library in Trinity College, Dublin is one measure of this. By 1836 the library had acquired a total of 78,686 volumes which had risen to 135,592 in 1860 and by 1900 there were 262,048 in the collection. Individual scholars likewise accumulated libraries on a scale not before seen while at a more popular level the rise of public libraries from the 1850s placed volumes in the hands of those who could not afford to purchase them.

The importance of print did not lessen with the new century. New publishers, such as the Talbot Press, emerged to meet the needs of the twentieth century. The market for collectors of fine printing was also served by the Cuala Press and later by Liam Miller's Dolmen Press. Newspapers and literary magazines, such as *The Bell*, came to have an important role in shaping the cultural identity of the new Free State. It is in this world of print culture that the honorand of this volume, Michael Adams, has spent most of his life both as author and publisher. From Irish University Press to Four Courts Press he has, almost single-handedly, effected a dramatic change in the area of scholarly publishing in the humanities in Ireland. It is appropriate, therefore, that this volume of essays should tackle one of the fundamental issues in the study of the history of the book: the impact of print and publishing on the wider world and in particular its role in shaping intellectual life.

The dramatic expansion of printed matter in the nineteenth century was propelled by both changes in print technology on the supply side and growing literacy and affluence on the demand side. The scale of the growth in output should not blind us to the fact that print was already a well-established feature of Irish life before the nineteenth century. From the 1660s the Dublin presses churned out increasing quantities of books for domestic and foreign consumption. Nor should we be unaware that print was only one of the available means of communication. Through the seventeenth and eighteenth centuries social rules for the production and consumption of print became increasingly well-established. Andrew Carpenter's essay charts the various ways in which contemporaries used print. Groups of people wishing to circulate ideas, sometimes scurrilous, might do this by manuscript, print or speech, the choice being dictated by personal preferences or conventions surrounding this practice.

Increasingly, through the eighteenth century those who wished to communicate their ideas turned to print, a medium capable of reaching a wider group than manuscript or speech. Reading became an increasingly important means of intellectual engagement with worlds far beyond the familiar and parochial. Toby Barnard's essay on the library of one, otherwise obscure, Cork clergyman demonstrates how diverse and strange those worlds could be. James Hingston collected books related not only to his clerical profession but works of classical scholarship, law and history together with plays and novels for recreation.

The nineteenth-century world of print was, therefore, built on secure foundations. What differentiated the nineteenth-century experience from what had gone before was simply the greater volume of books, and hence contesting ideas, that was available to a wider reading public. Several essays in this volume isolate three vital areas in nineteenth- and early twentieth-century intellectual life on which print had a crucial impact: law, history, and literary culture. In each case the link between print and the evolution of ideas is explored through a case study which examines how particular authors used print to convince others of their own way of looking at society, past and present.

Almost from their origins in sixteenth-century Ireland, print and publishing were driven by the practice of law and the working of government. The practice of law required the sort of standardization and wide circulation of texts, whether parliamentary statutes or legal handbooks, that print could provide. The way in which the legal profession rose to the challenge of the proliferation of print in the nineteenth century is addressed by W.N. Osborough in his case study of the publishing activities of John Finlay. In the 1820s, Finlay turned his attention to publishing handbooks to make available works that reflected the quirks of Irish legal practice, rather than being derived from English exemplars. It was, he claimed, only by such treatises that the 'knowledge of its [the law's] rules and obligations can ever be made distinct and intelligible'. Relatively cheap print made the circulation of such books possible.

The second strand of nineteenth- and twentieth-century intellectual life that was shaped by print was the study of the past. Victorians took to the study of the past for many reasons and political motives were often as important as scholarly ones in determining publishing output in this area. Four essays in this volume concentrate on how the study of the past has been served by print and publication in nineteenth- and twentieth-century Ireland. Tadhg O'Keeffe concentrates on the archaeological world of H.G. Leask and the influences that shaped his approach to the study of Irish castles. The publication of Leask's major study *Irish castles* by Dundalgan Press in 1941, and its subsequent reprints, demonstrates the importance of the work of a small press in shaping the intellectual landscape of twentieth-century Ireland. The other three essays in this section deal with a rather different, and even more enduring, intellectual legacy of historical publishing in nineteenth-century Ireland. The expansion of print culture combined with scholarly advances in the study of documents made it possible to produce for the first time editions of seminal texts that are still used today. The institutional basis which made this possible for Latin and Irish texts is dealt with in Raymond Gillespie's essay which explores the involvement of bodies such as the Public Record Office, the Rolls Series and the Historical Manuscripts Commission in Ireland. The experience of editing and publishing individual texts with a Celtic background is treated in the contributions by Jean-Michel Picard and Bernadette Cunningham. These essays reveal the complex network of intellectual contacts that lay behind the final printed product, some within Ireland itself and some international. Both Reeves' edition of Adomnan's *Life of Columba* and O'Donovan's classic edition of the *Annals of the Four Masters* are still in use by scholars demonstrating not only the intellectual resources that underpin them but also the power of print to disseminate and preserve texts for use by later generations of scholars.

Finally, and perhaps most crucially, the world of literary culture was reshaped by the activities of publishers and the printing press. So extensive was this process that it is only possible to catch a glimpse of it in a volume such as this. For that reason the two essays reflecting on high culture and print examine areas not normally dealt with within the context of conventional criticism. Nicholas Allen's essay, for instance, reveals a hitherto unconsidered aspect of intellectual life in the early Free State: an alternative world revealed not in the solid world of scholarly print but rather that of the ephemeral magazines *Klaxon* and *Tomorrow* which survived for only a few issues in 1923–4. Such fleeting glimpses remind us that we are dealing not with a monolithic, culturally authorized, world of print but with many different strands. While print and publishing certainly had the power to preserve and transmit ideas, a concept central to most of the essays in this volume, they could at times be a subversive and fleeting experience. By contrast to *Klaxon* and *Tomorrow*, Thomas Moore's *Irish melodies* was at once popular and reflected both literary and musical orthodoxies. The world of print made it possible to follow Douglas Hyde's advice that 'Every home should have its copy of *Moore* and *Davis*', and its influence in shaping perceptions of Ireland's musical past was a powerful one. Yet, as Harry White's essay shows the reception of the *Melodies* and its influence on the understanding of the Irish musical past was rather different to what Moore might have intended. In the world of mass production of print the cultural messages that one received were not always those that were intended by the author.

The world of nineteenth- and twentieth-century print, therefore, was the result of interactions between ideas, technologies and readers. The medium through which it operated was the book or pamphlet and the production of books became an increasingly sophisticated business, exploiting mass markets and providing ways of circulating and preserving ideas. But it was not welcomed by all. There were those who reacted against the mass production of books, preferring to revert to an older, craft-based world. William Morris and his Kelmscott Press is the best-known example of this trend in an English context but there were also Irish examples. The Yeats sisters and their Cuala Press is one such example but another is the involvement of Sir Edward Sullivan in the world of bookbinding as examined by Nicola Gordon Bowe in her essay. Sullivan, a founder member of the Arts and Crafts Society of Ireland, was both a scholar of bookbindings and a bookbinder himself. As such he brought to his subject a practical knowledge that has given his work, not least his work on the Book of Kells, a permanent value for those who study the Irish booktrade.

A collection of essays such as this can only begin to penetrate the complexity of the print and publishing revolutions. There are many areas not explored here that would repay further investigation. However, taken together they demonstrate the importance of print and publishing in shaping and reflecting the intellectual life of Ireland in the past. As such these essays, written by authors who have been published by Michael Adams at Four Courts Press, are a fitting tribute to a scholar and publisher who has done so much to continue that process into the present.

Acknowledgments

UNLIKE MANY FESTSCHRIFTEN this volume does not mark a significant moment, such as a retirement, but is offered to celebrate the achievements of Michael Adams and as a mark of the respect and affection in which he is held by many in the scholarly community in Ireland and beyond. Its task, to honour Michael Adams for his contribution to modern print culture and intellectual life, has ensured that there has been a great fund of good will and a desire on the part of many people to be helpful in its production. First, our thanks must go to the authors of the essays for the way in which they have responded enthusiastically to the project and the demands of the editors. All are authors who have been published by Michael Adams over the years. The list of those who would have wished to contribute to the volume is much longer but space constraints mean that the authors of the essays here must stand as representatives of their own disciplines and they have borne that task uncomplainingly. Secondly, we have been fortunate in our publisher, Terri McDonnell of The Woodfield Press. Producing a scholarly work in Ireland today without the direct involvement of Michael Adams is a difficult task, but one that The Woodfield Press have willingly taken on. The choice of publisher is all the more serendipitous since Terri began her career in publishing under the watchful eye of Michael Adams. Thirdly, a number of institutions that have been associated with Michael Adams in publishing ventures over the years have also assisted and we are glad to acknowledge the cooperation of the National Library of Ireland, the Royal Irish Academy and the Representative Church Body

Library in making illustrations available for this volume. The book was indexed by Gloria Greenwood, and Martin Healy ably assisted Brad Morrow in compiling the appendix. Many others have helped to make this book possible and to all those concerned, including Robert Towers and Anthony Tierney who were involved in its planning, our best thanks are due for what has been a genuinely cooperative effort.

Contributors

JEREMY ADDIS is the editor of *Books Ireland.*

NICHOLAS ALLEN, Assistant Professor of English, University of North Carolina, Chapel Hill, is the author of *George Russell (AE) and the New Ireland, 1905–30* (2003), and editor, with Aaron Kelly, of *The Cities of Belfast* (2003).

TOBY BARNARD, fellow and tutor, Hertford College, Oxford, is the author of numerous works on seventeenth- and eighteenth-century Ireland including *Irish Protestant ascents and descents, 1641–1779* (2003) and *A guide to sources for the history of material culture in Ireland, 1500–2000* (2005).

ANDREW CARPENTER is Associate Professor of English at University College Dublin and Head of the UCD School of English and Drama. He is the editor of *Verse in English from Tudor and Stuart Ireland* (2003) and *Verse in English from eighteenth-century Ireland* (1998), and of modern editions of two books by John Dunton, *The Dublin scuffle* (2000) and *Teague land: or, A merry ramble to the wild Irish* (2003).

BERNADETTE CUNNINGHAM, deputy librarian, Royal Irish Academy, is the author of *The world of Geoffrey Keating: history, myth and religion in seventeenth-century Ireland* (2000).

MARTIN FANNING is editor at Four Courts Press.

RAYMOND GILLESPIE, Associate Professor of History, National University of Ireland Maynooth, is the author of a number of books on early modern Ireland and series editor of Maynooth Studies in Local History.

NICOLA GORDON BOWE is a Lecturer in the History of Design and the Applied Arts, NCAD, Dublin (National University of Ireland). Her books include *The life and work of Harry Clarke* (1989) and (with E.S. Cumming) *The Arts and Crafts Movements in Dublin and Edinburgh* (1998).

TADHG O'KEEFFE, Senior Lecturer, School of Archaeology, University College Dublin, is the author of *Romanesque Ireland: architecture and ideology in the twelfth century* (2003) and co-editor of *The manor in medieval and early modern Ireland* (2004).

W.N. OSBOROUGH, emeritus Professor of Law, University College Dublin, is the author of *Law and the emergence of modern Dublin* (1996) and *Studies in Irish legal history* (1999). He has also edited several books in the Irish Legal History Series, including, most recently, *Adventures of the law: proceedings of the 16th British Legal History Conference* (2005).

JEAN-MICHEL PICARD lectures in modern and medieval French at University College Dublin. He is the editor of *Aquitaine and Ireland in the Middle Ages* (1994) and co-editor of *Ogma: essays in Celtic studies in honour of Próinséas Ní Chatháin* (2001).

HARRY WHITE is Professor of Music, University College Dublin, and President of the Society for Musicology in Ireland. His most recent book is *The progress of music in Ireland* (2005).

Michael Adams: a brief appreciation

JEREMY ADDIS

MICHAEL ADAMS EARNED his first doctorate with his research for *Censorship: the Irish experience* (1968), which is still the reliable main resource for students of that aspect of Irish print culture. At the 2005 winter commencements at Trinity College Dublin he was awarded his second, the degree of Doctor in Letters, *honoris causa*, for his work at Four Courts Press which, as the citation said, has transformed the life of learning led by many scholars of this island.

Michael learned the business of publishing in the 1960s with the Scepter Press, and then with the newly-established Irish University Press, in which his qualities of modesty and diplomacy did much to mitigate some of that company's worst follies – of greed as well as grandiloquence. He knew, too, the value of the books IUP had produced, and on its collapse in 1984 was one of a consortium of publishers that bought the stock and formed Irish Academic Press (IAP) to sell it, and to complete some of the firm's unfinished projects.

In that thankless mopping-up operation Michael showed and exercised remarkable talents, that were to transform a chore into the significant enterprise that we celebrate in the present volume. An academic publisher must be something of a polymath. He must love and understand books and must also be a careful business manager. He must motivate staff, communicate enthusiasm and new ideas to them and to potential buyers, and at the same time exercise the banker's qualities of caution

Michael Adams at the winter commencements, 16 December 2005, Trinity College, Dublin, photograph by Robert Towers

and restraint. He must bow to the expertise of scholars, but at the same time have the knowledge and confidence to impose his own standards upon their work.

All this and more Michael Adams gave in generous measure to IAP and to another creation which soon followed the first: the Round Hall Press, that discovered and supplied a demand for books on Irish law and that sparked off lively competition from long-established English law publishers, and was eventually sold to one of them. Michael also found time to establish his own imprint, Four Courts Press. To understand why his own publishing imprint was needed, one must recognize another quality of the man: devotion.

In the inchoate, featureless agnosticism that seems to characterize our times, the great unquestioning motivation of the man who knows where he comes from and knows where he is going – the man who fears God – is often sadly lacking, leaving little by way of overarching motivation but the all too often unsatisfactory Mammon. Michael Adams as a member of the Opus Dei prelature is devout in the old-fashioned sense. One of his less known achievements was to translate into English the scripture commentary of the 20-volume Navarre Bible, and it was as a vehicle for this and for works of theology that Four Courts Press was started.

After disengaging from IAP at the end of 1995 – twenty years after its foundation – he was free from its demands and constraints. From theology and biblical exegesis to Celtic and Medieval studies had seemed a natural progression for Four Courts Press, and the further development from these to encompass almost every discipline in the humanities, as illustrated in Martin Healy's appendix in this book, was inevitable too, given Michael's ability to say 'yes' to works that deserved the oxygen of print.

Quoting Horace on those who were unwept and unknown for lack of a sacred bard, Trinity's citation remarks that without a good publisher the most brilliant writer is condemned to the long night of our ignorance. The publisher is too often seen by scholars as a more or less compliant and complaisant functionary, and it is true that much of his work is necessarily a finicky balancing of variables, from the size and spacing of type and the weight and bulk of paper to more critical problems of print-runs and prices. Every such detail can make or mar a book's success, but it is the publisher's ability at the same time to see the bigger picture – to understand the significance of a book within its discipline and perhaps also in the wider interdisciplinary conspectus – that marks him and his imprint.

From this it is clear that the in-house academic publisher – the university or college press – is at a disadvantage in competition with a 'free house' such as Four Courts, for the 'tied' press (to borrow the usage of pubs and brewers) is necessarily the servant of the faculties and of the greater institution. But success must always hang upon the book itself. That success may and often does redound to the glory of the institution and the faculty as well as to the writer, but things do not seem to work so well the other way around.

The technicalities of printing of 550 years ago set a terrible conundrum for publishers: the more the cheaper. As Michael Adams has had good reason to know – though he has never fallen for the fallacy himself – publishers have often mistaken this for an excuse to fill warehouses with unsellable books. Curiously it is only in the past decade or so that new technologies have begun to ease the problem and allow economical short runs, and so to enrich and enlarge available scholarship. And curiouser yet, the availability on the internet of information and opinion – I carefully don't say scholarship for so much else is admixed – electronically cross-referenced and indexed to the nth degree and searchable in nanoseconds, is having the effect of increasing reliance on books. If, as I suspect, it is the discrimination, experience and skills of publishers that are found lacking in the ethereal medium and attractively reliable in paper and print, all the more reasons to celebrate the life and work of Michael Adams.

Circulating ideas

Coteries, groups and the circulation of verse in English in early modern Ireland

ANDREW CARPENTER

IN RECENT YEARS, it has become clear that a considerable amount of verse in the English language was written, and circulated, in early modern Ireland.[1] There are indications that literary coteries – groups of like-minded individuals who passed to each other the verses they had written, sometimes amending or adding to the texts – flourished in various parts of the country and, especially, in Dublin Castle from time to time during the latter half of the sixteenth century and for most of the seventeenth century. The verse the members of these coteries sent to each other was usually 'occasional' verse – created for a specific group or for an individual member of the group – and it normally remained in manuscript, though in some cases (those of Katherine Philips and Jonathan Swift, for instance) the poems were later printed.

Much of the surviving early printed verse connected with Ireland – that produced in Dublin, Cork or London before *c.*1680 – seems to have been custom-printed in very small print-runs and so intended, not for general circulation to miscellaneous purchasers, but rather to be passed around closed groups or sent to specific recipients. Normally, printed verse survives the ravages of time better than verse in manuscript, but

1 See Andrew Carpenter, *Verse in English from Tudor and Stuart Ireland* (Cork, 2003).

very few poems printed in (or connected with) Ireland before 1680 exist today in more than a single copy. Prose printed in Ireland before 1680 is nothing like so rare though, since many of these prose works were historical, theological or polemical works, they were bound at the time of purchase and treated with care by their owners.[2] But so few examples of printed verse connected with Ireland have survived from this early period that one wonders whether the material we have now was printed for general distribution at all.

Other verse, particularly ballads, circulated widely in early modern Ireland, of course, though most ballads and chapbooks were imported from England or Scotland, and only one Irish-printed ballad – considered below – survives from before the 1670s. In addition, there must have been hundreds of indigenous popular songs circulating orally in the communities of English-speaking sixteenth- and seventeenth-century Ireland, though only a few of these – three of them connected with sixteenth- and seventeenth-century Fingal – have been preserved in written form.[3] We know that popular verse was also composed and printed in Ireland; for instance, the execution of John Atherton, bishop of Waterford, in 1640 provoked 'scandalous rhyming pamphlets' in Dublin,[4] but we have no record of any of them. The printed verse which has survived has done so because, in almost every case, it was bound into volumes before or soon after purchase and because the library in which it was deposited has, itself, survived. Unbound broadsheets and folded sheets are a lot more fragile and there must have been many copies of such Irish verse printings which passed round the coffee houses of late-seventeenth-century Dublin and Cork and, having served their turn, were recycled for the lining of pie dishes, for the lighting of pipes or, as Mary Davys put it, for 'clapping to Posterior'.

2 On the early book trade in Ireland, see three articles by Raymond Gillespie: 'The book trade in southern Ireland 1590–1640' in Gerard Long (ed.), *Books beyond the Pale: aspects of the provincial book trade in Ireland before 1850* (Dublin, 1996), pp. 1–17; the same author's 'Irish printing in the early seventeenth century', *Irish Economic and Social History* 15 (1988), 81–8; and his 'The circulation of print in seventeenth-century Ireland' *Studia Hibernica* 29 (1998), 31–58.

3 For more on this subject, see Hugh Shields, *Narrative singing in Ireland* (Dublin, 1993). The three Fingal texts are in Carpenter, *Verse in English*, pp. 58, 210 and 310.

4 Gillespie 'Circulation of print', 45.

However, material in manuscript is, and was also in the early modern period, regarded as potentially worth keeping, and a fair quantity of manuscript verse from early modern Ireland is still extant. It falls into two categories: 'polite' and 'impolite' or subversive verse. Polite verse could be addressed to a specific individual, such as the fifty or so verses sent to the first duke of Ormond during the periods he served as viceroy in Ireland, or the verse letters sent by Lady Anne Southwell of Pounelong Castle in County Cork to the bishop of Limerick and others in the 1620s – though we only know of the latter from Lady Anne's commonplace book where she drafted the verses which we presume that she sent.[5] 'Polite' is a term which can be used also of the earliest example in modern English of a verse epistle sent from one group in Ireland to another group in Ireland; this was the letter sent, in 1487, by the mayor and inhabitants of Waterford to Walter Fitz-Simon, the archbishop of Dublin, and to 'the mayor and citizens of the same, in the tyme of their rebellion'. The senders hoped that this long poem would discourage the archbishop and citizens of Dublin from supporting the claim to the throne of Lambert Simnel, but its importance from our point of view is that it shows that one group of citizens in early modern Ireland chose to communicate with another group in verse. Since verse was used regularly in the sixteenth and seventeenth centuries to put forward a political point of view in print, it is only reasonable to assume that it was used to persuade or to browbeat in manuscripts now lost.[6] Other early texts intended for a 'polite' environment include an Elizabethan court poem by Richard Bourke, fourth earl of Clanrickard, who praised the 'beames and joyes' in the eyes of the elderly Queen Elizabeth. There must have been many other examples of courtly, political or social verse of Irish origin from this early period which have not survived.

5 The majority of the letters addressed to Ormond are in a volume in the Beinecke Library at Yale University (Osborn MS Fb 228); Lady Anne Southwell's commonplace book is MS V.b. 198 in the Folger Shakespeare Library, Washington, DC.

6 The Waterford letter survives in a sixteenth-century transcription by Meredith Hanmer, historian of Ireland and rector of the College of the Blessed Mary at Youghal, County Cork. The National Archives, Public Record Office, London, SP.63/214.

There is enough evidence, however, for us to deduce that life at the centre of power in sixteenth- and seventeenth-century Ireland, Dublin Castle, was stable enough to allow a limited level of literary activity. We find records of verse being composed and circulated in and around the vice-regal court in Dublin several times between the governorship of Sir Henry Sidney in the 1550s and that of Lord Berkeley in the 1690s. Some viceroys – Sir Thomas Wentworth and the duke of Ormond, for instance – went out of their way to try and create in Dublin Castle a literary culture similar to that which flourished in the courts of Renaissance Europe and those of the Tudor and Stuart monarchs. But their successes were short-lived – Ireland was a rough and 'barbarous' place – and nothing like the body of literary work which graced some foreign courts ever developed in Ireland. As Raymond Gillespie notes wryly 'there were few figures of literary merit' in early modern Ireland.[7] But there were, in each generation, people who wrote verse which they circulated to their friends; in Dublin Castle, people who had known each other in England were thrown together and made new friends, and in urban centres such as Cork and Dublin, coteries were formed by those who shared political or intellectual interests. Other, looser groupings or coteries existed in the countryside, or linked poets in Ireland with those in England. In some cases, members of such groups passed to each other copies of 'naughty', iconoclastic or subversive verse – as did the groups of which Sir William Petty and Sir Hans Sloane were members in the late seventeenth century.[8] In other cases, individuals would circulate to each other witty poems designed to tickle each others' fancy or to congratulate or commiserate with each other.[9] It is ironic that this kind of verse, intend-

7 Gillespie 'Irish printing', 85.

8 The Sloane collection contains several coarse poems set in Restoration Ireland (Carpenter, *Verse in English*, pp. 382–401) and the British Library manuscript volume entitled 'Sir William Petty's poetical amusements' contains, *inter alia*, a verse 'Letter from a Missionary Bawd in Dublin, to her chief [*sic*] in London' (Carpenter, *Verse in English*, pp. 423–7).

9 For an example of the witty verses sent to each other by doctors practising in Cork city in the late 1660s, see 'On the praise, and happy delivery of James Wolveridge, Dr. of Physick, in his labours on the labour of women &c. delivered by Aquila Smyth, MD' in Carpenter, *Verse in English*, pp. 418–20.

ed for a coterie or a strictly limited group of friends, has survived, while almost all the verse intended for wide circulation in early modern Ireland has disappeared without trace.

At the same time, we can be sure that books of verse from Renaissance Europe and from England were read in professional and legal circles and in the upper echelons of polite society in Ireland, though much of it may have been imported privately. When, in the 1590s, the young James Ussher, later archbishop of Armagh, declared that he 'always loved a good poem himself that was well and chastly write', he was issuing a judgment based, presumably, on considerable experience of verse of various kinds.[10] Similarly, when Edmund Campion praised the quality of the intellectual life he experienced while staying at the house of James Stanihurst in Dublin in the 1570s, he was comparing it with a city like Oxford where verse was readily available.[11] We also know that substantial quantities of books were imported into Ireland in the sixteenth and seventeenth centuries, at least some of which must have been books of Renaissance poetry – though surviving lists of imports do not indicate books of verse as such.[12] Sir John Harington distributed copies of his own translation of *Orlando Furioso* in Ulster and saw the book being read in Galway.[13]

The most tantalizing sighting is of what appears to have been a quintessentially Renaissance piece of writing originating in Ireland. In 1934, a fragment of an early Dublin printing of a verse rendering of the story of the destruction of Troy – taken from a binding – was shown by the

10 Richard Parr, *The life of … James Usher, late lord archbishop of Armagh* (London, 1686), pp. 3–4. See also Raymond Gillespie, 'The social thought of Richard Bellings' in Micheál Ó Siochrú (ed.), *Kingdoms in crisis: Ireland in the 1640s* (Dublin, 2001), pp. 212–28 (p. 213).

11 Edmund Campion, *Opuscula Omnia*, ed. R. Turner (London, 1625), p. 208 and Edmund Campion, *Two bokes of the histories of Ireland*, ed. A.F. Vossen (Assen, 1963), pp. [6], [7] and [108]. Quoted in Colm Lennon, *Richard Stanihurst the Dubliner, 1547–1618* (Dublin, 1981), p. 28.

12 See for example the lists in Gillespie 'Irish printing', 87–8 and Gillespie 'The book trade', 13. In the second of these, 201 of the 2675 books listed are in a category simply labelled 'Books'.

13 John Harington, *Nugae antiquae*, ed. Henry Harington, 2 vols (London, 1769), i, p. 340.

London bookseller, Frank Marcham, to the American bibliographer W.A. Jackson. The fragment, which has since disappeared, contained the information that it had been printed in Dublin for a Francis Edderman, presumably the poet himself. It also indicated that the poem had been dedicated to Sir Henry Sidney, and the printing was therefore tentatively dated to the year 1558, a time when Sidney was in Dublin during his period as lord justice. Even if we did not have this glimpse of a Dublin-printed Renaissance poem dedicated to Sidney, we could plausibly surmise that Sidney, whose children Sir Philip Sidney and Mary, countess of Pembroke, became the stars at the centre of the most famous literary coteries of the time, would have tried to make his court in Ireland a place where literary coteries might flourish.[14] The violence of the times frustrated any long-term plans Sidney might have had, but the Edderman fragment suggests that there was some kind of literary life in Dublin in the 1550s and that the lord justice was prepared to support it. Several other Irish works were dedicated to members of the Sidney family at this time,[15] and one wonders whether Edderman was seeking Sidney's patronage so that he could become part of a vice-regal literary coterie.[16]

When Edmund Spenser was in Dublin in the 1580s as secretary to the lord deputy, Lord Grey, he became part of a circle of cultivated men living in the city at the time who were in the habit of meeting to discuss ideas and literary matters. A particular friend of Spenser was a young Italian named Ludowick Bryskett, a protégé of Sir Henry Sidney and close companion of his son, Sir Philip Sidney. Bryskett wrote two fine elegies on the death of Sir Philip Sidney which appeared in print with Spenser's own elegy on Philip Sidney in 1595. The circle in which Spenser and Bryskett moved in Dublin was similar to that which had earlier sur-

14 See H.R. Woudhuysen, *Sir Philip Sidney and the circulation of manuscripts, 1558–1640* (Oxford, 1996), pp. 196, 207–23.

15 Anne Fogarty, 'Literature in English, 1550–1690: from the Elizabethan settlement to the Battle of the Boyne' in Margaret Kelleher and Philip O'Leary (eds), *The Cambridge history of Irish literature*, 2 vols (Cambridge, 2006), i, pp. 140–90.

16 Though Raymond Gillespie is, of course, correct when he writes of 'the lack of a court society in Ireland to match that in England where literary and other artistic patronage was encouraged' (Gillespie 'Irish printing', 85), the evidence suggests that attempts were made, from time to time, to create such a society.

rounded Sir Philip Sidney.[17] Bryskett gives a graphic account of a meeting of the group in his translation of the second part of Giraldi's *De gli hacatommithi* which he published as *A discourse of civill life.* This work is cast in the typical Renaissance form of a series of dialogues between three or four gentlemen, during which the text is presented and discussed. Bryskett stated that 'the occasion of the discourse grew by the visitation of certaine gentlemen comming to me to my little cottage which I had newly built neare unto Dublin'[18] and he printed his translation as if it had been the subject of discussion between his friends at a three-day gathering at his cottage. Among the company, he writes, were 'Doctor *Long* Primate of *Ardmagh*, Sir *Robert Dillon* Knight, M. *Dormer* the Queenes Sollicitor, Capt. *Christopher Carleil*, Capt. *Thomas Norreis*, Capt *Warham St Leger*, Capt. *Nicholas Dawtry*, & Mr *Edmond Spenser*, late your Lordships Secretary, & *Th. Smith*, Apothecary.'[19] When Spenser is invited by Bryskett to deliver to the gathering a dissertation on moral philosophy, he declines and suggests that Bryskett's own translation of Giraldi would be the ideal text to be read to the company. But Spenser also reveals something of the process of the writing of *The Faerie Queene* in his reply:

> For sure I am that it is not unknowne unto you, that I have already undertaken a work tending to the same effect, which is *heroical verse*, under the title of a *Faerie Queene*, to represent all the moral vertues, assigning to every vertue, a Knight to be the patron and defender of the same: in whose actions and feates of armes and chivalry, the operations of that vertue, whereof he is the protector, are to be expressed, and the vices & unruly

17 The volume in which these elegies appeared was *Astrophel* – which Danielle Clarke has surmised may be the result of a poetic jousting which could have taken place in Ireland some time during the nine years between Sidney's death in 1586 and the publication of the book in 1595.

18 Ludowick Bryskett, *A discourse of civill life: containing the ethicke part of morall philosophie. fit for the instruction of a gentleman in the course of a vertuous life* (London, 1606 [facsimile London, 1972]), p. 5.

19 Bryskett, *Discourse*, p. 6. The person to whom the whole document was addressed ('your Lordship') was Lord Grey. See Pauline Henry, *Spenser in Ireland* (Cork, 1928), p. 43.

> appetites that oppose themselves against the same, to be beaten downe and overcome. Which work, as I have already well entred into, if God shall please to spare me life that I may finish it according to my mind, your wish (M. *Bryskett*) will be in some sort accomplished ...[20]

Though some doubt has been cast on the accuracy of the details of Bryskett's account,[21] there is no reason to question the essential truth of his story of a group of well-educated young men living in or near Dublin in the late sixteenth century engaging in the exchange of ideas and comments about literature and reading to each other their latest verses. What is worth noting here is that all those listed were either working in, or would often have been guests at, Dublin Castle. Passages at the beginning of Spenser's *Colin Clouts come home againe* suggest that a similar coterie formed around him and Sir Walter Ralegh when Spenser moved to his newly acquired estates at Kilcolman, County Cork, in the mid-1580s.[22] Colin, the poet, recites parts of the poem to his fellow shepherds (including the 'shepherd of the Ocean', Sir Walter Ralegh); they respond to him and, even if the pastoral details of the scene are clearly fanciful, there is more than a suggestion, once again, of a verse coterie active in early modern Ireland.[23] The verses written by Spenser, Bryskett and others living in Ireland were all printed in London where, as Gillespie puts it, they would find 'a more appreciative audience'.[24]

Richard Nugent of County Westmeath was another Irish poet whose work appeared in London. His sonnet sequence, *Cynthia*, which was printed in 1604, contains not only poems in which Nugent bemoans the loss of his Irish mistress and his need to move from Westmeath to the 'uncouth land' of England, but also a sonnet addressed to his cousin

20 Bryskett, *Discourse*, pp. 26–7.

21 The earliest doubter was Professor John Erskine whose article (*Proceedings of the Modern Language Association*, 30 (N.S., 23, 1915), 831–50 has been discussed and rebutted by several more recent scholars. See also Henry R. Plomer and Tom Peete Cross, *The life and correspondence of Ludowick Bryskett* (Chicago, 1927), pp. 77–83.

22 *Colin Clouts come home againe*, line 17.

23 *Astrophel* contains poems by three writers connected with Ireland, Spenser, Bryskett and Sir Walter Ralegh.

24 Gillespie, 'Irish printing', 86.

Richard Nugent of Donower, and a poetic answer from him. If we are to believe the story as it unfolds in Nugent's book – and, again, there is no particular reason not to do so – he and his Westmeath cousin were writing verses to each other and exchanging them as a small coterie. By printing his sonnet sequence, Nugent invites a wider audience to enjoy his work; but again, the fact that only a single copy of this little book survives suggests very limited distribution.

Evidence of poetic and literary activity in or around Dublin Castle surfaces again in 1610. The English poet, George Wither, is thought to have spent some time in Ireland during his early manhood, when he was almost certainly attached to the household of the Loftus family at Rathfarnham Castle, near Dublin. During this time, he composed an extraordinary poetic masque – unique of its kind in Ireland. From this text and another contemporary description of the event, we learn that a large contingent of nobles, soldiers and their ladies, led by the lord deputy, Sir Arthur Chichester (later Baron Chichester of Belfast), came to Rathfarnham from Dublin Castle to celebrate the marriage of Lady Cassandra Ridgeway and Sir Francis Willoughby in October 1610.[25] The text specifically addresses not only the viceroy himself and other nobles, but the 'brave Martialists' who are fresh in from skirmishes in the bogs; the masque itself was intended to provide a short period of peaceful fantasy in the midst of an unsettled and warlike world. The poet assures the lord deputy that 'all are good friends' here and that it is safe for him to play, dance and sing since his enemies are 'quelled'. The stage directions attached to the text indicate a pause to allow the assembled nobility to listen to 'heavenly music' and, a few moments later, a page – styled 'lovely Ganimead' – appears bearing wine 'in a cup of massy gold'. The text contains many suggestions that what is going on in the performance is paralleled by the life of the classical gods, and builds up a dream-like

25 There is a long and vivid description of the ceremony – mentioning 'maskings, feastings, fireworks and presents' – in a letter sent by Thomas Ridgeway, father of the bride, to the parents of the groom, Sir Percival and Lady Willoughby. See David Cressy, *Birth, marriage, and death: ritual, religion and the life-cycle in Tudor and Stuart England* (Oxford, 1997), pp. 290–1. The letter is in the Huntington Library, San Marino Stowe MSS STB Box 2(1). My thanks to Danielle Clarke for this reference.

atmosphere typical of English court masques. This provides strong evidence for the existence of at least a rudimentary court culture, in which poetry, singing, dance and performance are found, in the circle surrounding the lord deputy in early seventeenth-century Ireland. Though Wither's masque is a poem – it is in rhyming iambic pentameters – it may not be a coterie poem as such; but it does reflect a world in which poetic coteries almost certainly existed. Like many texts originating in aristocratic circles or in coteries, Wither's masque remained in manuscript in the archives of a noble family until two hundred years after its first performance.[26]

A text which is surprising to find in contemporary printed form is the volume of verse produced in Dublin by members of Trinity College on the death of one of its patrons, the countess of Cork (*Musarum lachrymæ*, Dublin 1630). Such material is normally found in a single, presentation manuscript, not in print – though again, since only two copies of this book survive, it is likely that only a few were printed and that the volume never received wide circulation. One could speculate that the poems which make up this volume were put into print because the Trinity College poets concerned wanted copies for themselves; but even if this is so, it seems likely that the book was produced in a print-run of a handful of copies; it is, incidentally, surprising that no copy remains in Trinity College itself. The volume is a significant survival, however, since it shows Dublin printers of the reign of Charles I able to set and print many different fonts, including Greek and Hebrew. Few outside Trinity College could have read these texts, and it is virtually certain that *Musarum lachrymæ* was produced for circulation only in the college and in the Boyle family. This printing is of a coterie text that has escaped into print. And it seems beyond question, also, that texts such as those I have been discussing existed and circulated in other Irish coteries of which we now have no evidence. Sometimes, for reasons which remain obscure, the poems might have been printed, but only for circulation within the group.

26 It is printed in *The continuation of the history of the Willoughbv Family* by Cassandra, duchess of Chandos (1679–1735), ed. A.C. Wood (Eton, 1958), pp. 82–4, and in Carpenter, *Verse in English*, pp. 135–8.

An example of early Irish-printed verse which, rather surprisingly, survives in only a single copy is the collection of poems by Richard Bellings at the end of the Dublin 1624 printing of his *Sixth booke to the countess of Pembroke's Arcadia*. One would have expected Bellings' additional book to the *Arcadia* to have attracted a reasonably large readership – like anything connected with Sir Philip Sidney, himself the most famous coterie poet of all. The book is a substantial quarto to which Bellings' poems are appended with the following preliminary statement:

> THESE FOLLOWING VERSES at severall times came to the hands of the right worth[y] Sir R.C. who being the Authors deere friend, & therfore thinking them too good to perish, hath caus'd them to be annexed to his booke.

This describes exactly how verse circulating in manuscript in a coterie can appear in print, when the timing is right and the opportunity presents itself. Bellings was a student at Lincoln's Inn in London when he undertook his additional book of the *Arcadia*, and his coterie, including 'Sir R.C.', must have existed in the London inns of court – through which so many young seventeenth-century Irishmen passed. The coterie is likely to have been made up, at least to some extent, of Bellings' countrymen, and some of his poems have direct links with Ireland; one is a vivid 'Description of a tempest' at the end of which 'The wind is calm'd, and we at Skerries land'. Members of Bellings' London coterie included W. Martyn (probably a member of the prominent Catholic family of the Martins of Galway) and W. Delaune, both of whom contributed commendatory verses to Bellings' poems. Unfortunately, in the sole surviving copy of this book, someone has vigorously erased (until it is illegible) the name of the author of the most interesting of the dedicatory verses, that entitled 'TO HIS ENTIRELY BELOVED Kinsman the AVTHOR'. That poem praises Bellings for his work, 'Which benefite, thy country and thy friends / Reape from the happie labours of thy youth …' The identity of this member of the coterie is, for the moment, unknown.

Readers for this printing of Bellings' work were clearly expected not only among courtly followers of the cult of Sir Philip Sidney but also among Bellings' London-Irish coterie. The reason that this courtly romance, destined primarily for a London audience, was printed in Dublin is that it enabled the Dublin Society of Stationers to use printing-presses which would otherwise have stood idle. In other words, the decision to print this book in Dublin was a commercial rather than a cultural one.[27] However, that some courtly audience was expected in Dublin as well as in London is indicated by the fact that Bellings dedicated the text to Elizabeth Cary, Countess Falkland, wife of the Irish lord deputy. Lady Falkland was a fine poet herself and one of the most learned women of her time; it would be extraordinary if she had not encouraged around her in Dublin Castle a circle of courtly readers of verse – and possibly of writers of verse.[28] Bellings is said to have taught Lady Falkland the Irish language. Though we have no evidence for it, the supposition that there were poetic coteries flourishing in Dublin Castle in the 1620s is too plausible to dismiss.

Lord Deputy Falkland was the man whose actions created the context for the only surviving Dublin-printed broadsheet of the early seventeenth century, *Mount Taraghs triumph*. Though the text carries a statement that it is to be sung 'To the tune of the Careere',[29] it seems unlikely that the verse was designed for performance in the street or the tavern. The verse is full of ironies which would be lost in street performance; the images are sophisticated and the vocabulary nuanced. The third stanza, for instance, seems dangerously close to parody; troops loyal to King Charles have been summoned to appear not at Dublin Castle but at the hill of Tara, the ancient seat of the high kings and the venue for the legislative assemblies of pre-conquest Ireland.

> The Squadrons appeere, the Troupes they Careere,
> from the foote of the Hill, to the top:

27 Gillespie 'Irish printing', 85.

28 Two of Lady Falkland's poems and a useful short biography are in Jane Stevenson and Peter Davidson (eds), *Early modern women poets: an anthology* (Oxford, 2001), pp. 143–6.

29 A tune written by William Basse (d. *c.*1653) entitled either 'Maister Basse his Careere' or 'The Hunter in his Careere' was current in London (and perhaps in Dublin) in the 1620s.

With force (like a flood, being mainely[30] withstood)
their courage indureth no stop.
King *Charles* to thy Troupes, the hill of *Taragh* stoupes
that long a Royall Mansion hath beene:
With Ecchoes in the skies, it shouts and it cries,
long life to King *Charles* and his *Queene*.

These verses, ostensibly attempting to urge the Catholic Old English to support Charles I's Irish administration, seem more designed to provoke mirth in a circle of cynical New English parliamentarians than loyalty in the hearts of the Old English. Though it might look like a 'public' piece, it is quite possible that *Mount Taraghs triumph* was designed to be circulated in a coterie of like-minded individuals where its ironies could be savoured in private rather than sung in the street.

Work needs to be done on another group of poets based in seventeenth-century Ireland – the group surrounding Thomas Wentworth, earl of Strafford, during his term as viceroy in Dublin in the 1630s. The main evidence for the poetic activities of this group is found in the prefatory verses affixed to various printings of the works of James Shirley, the playwright who came from London to provide plays for the Dublin stage.[31] Among the members of the group who wrote poems praising Shirley and commending the court in which they were all living were Richard Bellings, Andrew Cooper, and William Smith – three poets with other published verse to their names. The poems indicate clearly that there was an active group of poets in Strafford's court – a court which came to an abrupt end with the impeachment of Strafford and the 1641 rising. But, from our point of view, this is evidence of another poetic coterie based in the Irish vice-regal court; once again, however, the place where this material reached print was London.

Irish-printed texts which prove the existence of seventeenth-century coteries outside Dublin include two poetic squibs from Cromwellian

30 with vigour.

31 These are easiest to access in the preface to *The dramatic works and poems of James Shirley* ed. William Gifford and Alexander Dyce, 6 vols (London, 1833, reprint New York, 1966).

Cork, William Wheeler's 'To the Honorable Commissioners for Assesments, the complaint of the south suburbs of Corke' and William Wright's 'A Preparative to a pacification betwixt the south and North Suburbs of Corke' – both printed in Cork in 1656. These poems were written to tickle the humour of the citizens of Cork at a time when the city was attempting to regain some measure of self-governance from the Commonwealth military authorities in Dublin. The poems are full of references to individuals and places which, though impossible to identify accurately today, show that the texts were designed for reading and circulation in a small, defined group – a coterie. One poem directly answers the other but it is not clear why they were printed at all since the circle of those able to understand them must have been a small one. Only single copies of each text remain today.[32]

Another early municipal poem is 'An Irish banquet, or the mayor's feast of Youghall', from 1604. The references in this poem are to individuals intimately known in the small circle of those deliberating in the council-chambers of the corporation of Youghal: this is not a coterie poem, as such, but a poem designed for distribution within a very limited circle. That this circle included not only the council members but also their wives and friends is suggested by the fact that the heroine of this wonderfully iconoclastic and bawdy poem is the wife of the incoming mayor. The superiority of the women in the poem is made clear as the reader is told to fear not the 'rebels' of the day but Catholic priests and the women of Youghal:

> Feare not *Tirconnel*, nor those *Galliglasses*[33]
> That cut, and hacke, and carve men as it passes;[34]

32 See Andrew Carpenter, 'Two early printed verse squibs from Cork' in Charles Benson and Siobhán Fitzpatrick (eds), '*That Woman!': studies in Irish bibliography: a festschrift for Mary 'Paul' Pollard* (Dublin, 2005), pp. 163–74.

33 Even though they had been defeated at the battle of Kinsale in 1601, O'Neill and O'Donnell posed at least a potential threat to English planters and settlers in Ireland until 1607 when they and their followers left Ireland in 'the flight of the earls'. Gallowglasses (Ir. *gallóglaich*) were heavily armed infantry, originally mercenaries from Scotland, who were integrated into the marauding bands of Tyrone, Tyrconnell and others, and were widely feared.

34 i.e. as they pass.

> Feare those which all these feare, those fathers holy[35]
> Which make the whole world their sole monopoly:
> That crowne & uncrowne Kings, when as they please,
> Play fast and loose like Juglers with slight ease;
> Dissolve all othes, though made with hand & heart,[36]
> And pardon all sinnes, yea an Irish fart.[37]
> Feare these, and those they joyne with,[38] lest too late
> We finde our Ile an Amazonian state ...

After 1660, the rate of survival of Irish-printed books of verse in English increases. In two cases, we learn that these are Irish printings of poems previously circulating in manuscript in Irish coteries or groups. The earlier example is Samuel Dancer's 1663 Dublin printing of *Poems by several persons of quality and refined wits*, containing several poems by Abraham Cowley and three by Katherine Philips. This Welsh poet was part of the poetic coterie surrounding the duke and duchess of Ormond in Dublin Castle in the early 1660s. The members of the circle – which included the earl and countess of Roscommon and several members of the Boyle and Butler families – gave each other coterie names (Philips was known in Dublin – as she had been before she came to Ireland – as 'Orinda', other coterie members as Celimena, Policrite and Silvander) and addressed verses to each other. Katherine Philips was understandably annoyed that her three poems – including one written for the marriage of Lady Mary Butler and William Cavendish, fourth earl (and later first duke) of Devonshire, an event which took place in Kilkenny in October 1662 – had reached the public by means of the (at that time in Dublin) unrestricted press, instead of remaining within the controlled circulation of the courtly coterie. But there was nothing she could do: her manuscript poems were now printed and so she asked her friend Sir

35 i.e. the pope and the cardinals in Rome.

36 A reference to the pope's excommunication of Queen Elizabeth in 1570.

37 *Marginal note*: 'The eighth deadly sin; and more shunned of the Irishmen than the other seven.' Throughout the early modern period, the extraordinary belief that Irishmen had a rooted objection to farting is regularly repeated in English accounts of Ireland.

38 *Marginal note*: 'Their crosses, their wives.'

Charles Cotterell to start collecting her poems for an authorized printing. When this came, in 1667, it was a lavish, folio production in which her work looked much more imposing than it had in John Dancer's hastily-printed *Poems by several persons.* But Dancer's text was an accurate one and it seems clear that, though the Dublin Castle poems were not intended to appear in print, a member of this 'polite', social coterie had simply allowed Dancer to take a copy of the poems. He, presumably, went to print in the hopes of making some money out of a Dublin reprinting of Cowley's poems, spiced up with some local material by 'Persons of quality'; in other words, Dancer expected to generate extra sales by implying that the purchaser could gain access to the otherwise closed world of a Dublin Castle coterie made up of 'the quality'. This seems to be the first book of poetry printed in Dublin which the bookseller expected to sell in his shop and from which he expected to make a profit. It is strange, again, that only a single copy of this publication has survived.

Another interesting case of transfer from manuscript to print is that of the 1675 Dublin printing of a verse translation of Juvenal's tenth satire. This translation was almost certainly the work of Dudley Loftus, though the title-page coyly refers to the author as 'A Person, sometimes Fellow of Trin[ity]. Col[lege]. Dublin'. The book is prefaced by a letter in which Edward Wetenhall, later bishop of Cork, dedicated the book to Viscount Blessington; Wetenhall describes the author as someone 'drawn to the Terms' (that is, the court sittings) in Clonmel by 'some perplext business'. As Tony Sweeney has pointed out, Dudley Loftus was the only ex-fellow of Trinity College, Dublin involved in legal affairs in Clonmel at this time – and he had the scholarship to have undertaken the translation of Juvenal.[39] Wetenhall's preface also includes interesting information concerning the publication of the poem, including a rebuke to Blessington for allowing what had been intended as a poem for private, limited circulation in manuscript, to escape into print.

39 Personal communication: but see also Tony Sweeney, *Ireland and the printed word: a short descriptive catalogue of early books, pamphlets, newsletters and broadsides relating to Ireland, printed 1475–1700* (Dublin, 1997), entry 2646, p. 445.

> Returning home [i.e. from Clonmel to Dublin], he [the translator] reviewed what he had done, and moved by the worth of the subject, gave three Copies to Persons of Quality, known by him to be addicted to Literature, (of whom your Lordship, by sending your Copy to the Press, not without great encouragement, made it known you were one) designing onely thereby, farther to invite them to the reading such old Books, which at once present much Wit, Learning and Morality. From the Press the Papers came to my hands; and I should neither do your Lordship nor my Friend justice, if from mine they did not return to yours.[40]

Blessington seems to have broken the convention that those to whom poems were circulated in manuscript should have kept them private – though it is not clear what the phrase 'not without great encouragement' means in this context. However, one wonders how many other poems, circulating in this way in Ireland at this time, never passed into print.

One particularly eccentric poet whose work, though intended for a small audience, has survived in Irish printings, is the ex-soldier William Mercer. His first Irish verse seems to be *A VVelcome in a poem to His Excellency John Lord Roberts Baron of Truro, Lord Lieutenant General, and General Governour of Ireland, my most noble patron &c. at his royal entry into the Castle of Dublin*' which was printed in Dublin in 1669. Mercer was clearly unsuccessful in his attempt to extract patronage from Lord Robartes (who was only lord lieutenant for nine months in 1669–70) and he declared himself unable to obtain an engraving of a likeness of Robartes as a frontispiece for his groveling poem. Instead, he printed an empty frame and invited the reader to fill it in as he would. The production of this piece of post-modern fancy must have been paid for by Mercer himself and the printed poem represents an attempt by one of the many poetasters of the day to ingratiate himself into the vice-regal circle. However, Mercer seems to have been the only poet brazen enough to put his sad offering into print. The separate title page at leaf A3r tells

40 'The Wish ...' (Dublin, 1675), sig B4.

the reader (of whom there must have been precious few) that this is a work in which the reader can hear 'Mercer's muse making melody'.

Mercer's Cork 1675 printing of his poem *The moderate cavalier or the soldier's description of Ireland and of the country disease* is equally peculiar. Mercer, who had somehow obtained an estate in Ireland after his time in the royalist army, intended to present three dozen copies of his book 'to the Mayor of Cork and his guests at his feast when he first comes into office'. Mercer addresses his intended targets as follows:

> Gallants, to crown this Feast, you're here presented
> With a new dish of Poetry, Invented
> By a Soldier, and brought from Camp to th'City
> VVich speaks of serious things more than witty …[41]

Though he does, at one place, suggest that he wants his book to be read by 'the Protestant nobility gentry and soldiery of Ireland', the volume as a whole shows Mercer trying to ingratiate himself into a circle where his work might be read. Here is a poet who wished he was a member of a select group but who had to make do with the printing press instead. After 1680, there is an increasing rate of survival of original Irish-printed verse – though it is still meagre in comparison with the riches available in English-printed material of the period. Less than twenty Dublin-printed broadsides survive from the 1680s and only a handful from the 1690s – though there is always the hope that other Irish-printed items may turn up as library cataloguing improves. Only a few years ago, the American scholar James Woolley located a unique copy (in the Derry and Raphoe Diocesan Library) of Jonathan Swift's 1690 Dublin-printed *Ode. To the king*.[42] There are also a few surviving volumes of verse printed for mass circulation – though again, the numbers known are very low. Into this category come a handful of surviving seventeenth-century Irish almanacs and the volume of hymns and religious poems

41 William Mercer, *The moderate Cavalier* (Cork, 1675), pp. 2–3.

42 James Woolley, 'Swift's first published poem: *Ode. To the king*', in Hermann J. Real and Helgard Stöver-Leidig (eds), *Reading Swift: papers from the fourth Münster Symposium on Jonathan* Swift (Munich, 2003), pp. 265–84.

written by Bishop Luke Wadding of Ferns and published, ostensibly in Ghent, in 1684. We learn from Wadding's diary that he distributed dozens of copies of his *Smale garland of pious and godly songs* to his flock, though only a single copy of this fragile item has survived. This reminds us of the widespread circulation of religious verse in Protestant as well as in Catholic circles in seventeenth-century Ireland;[43] once again, however, the circulation of this sectarian or proselytizing verse was limited to those with particular convictions and those they sought to convert. In a less pious vein, the medical doctors of Cork, a small and specialized group, waxed poetical in their praises of each others' skill in midwifery in the verses prefaced to James Wolveridge's *Speculum matricis Hybernicum, or, The Irish midwives handmaid* (London, 1670); this group of encomiastic poems shows the doctors writing poems to and for each other and circulating them in what we would now call a coterie. Once again, the resulting text was printed outside Ireland.

Members of coteries might hoard the verse they received themselves – as did Sir William Petty and Sir Hans Sloane (both of whose collections are now in the British Library) – or they might deposit them in the family archive, as did members of the Boyle and Willoughby families.[44] Impolite verse – both subversive and 'naughty' material – was also hoarded and is found today in various places, in the Sloane and Petty collections in London, for instance, and in the 'Whimsical medley' a verse miscellany assembled between the 1680s and the 1720s by Theophilus Butler, first Lord Newtown Butler, now in the library of his *alma mater*, Trinity College, Dublin. Among the hundreds of poems in the 'Whimsical Medley' are several scurrilous and indecent poems which were circulating, in manuscript, in Trinity College in the 1680s; there is also the full text of the 1688 Trinity 'tripos' speech – a complex, multi-sectioned text of the iconoclastic, anti-establishment speech delivered by the *terrae filius* – the student chosen to satirize the university lectur-

43 See Raymond Gillespie, *Devoted people: belief and religion in early modern Ireland* (Manchester, 1997), particularly pp. 153–8.

44 The Boyle family material is among the Orrery papers at Petworth House, Sussex, and the Willoughby material (including the transcript of the masque by George Wither) at the University of Nottingham.

ers and professors at the conferring or commencement ceremony. The 'speech' itself is made up of several sections, in prose, in verse and in dialogue: the sections vary considerably from each other and I have argued elsewhere that the 'tripos' as a whole is the production of a coterie – in this case made up of the senior students at Trinity College. I believe that, though a student named John Jones was the elected *terrae filius* who actually delivered the speech – or part of it – on the day of the commencement, the text itself had been constructed during the weeks or months beforehand by members of the senior class – a coterie.[45] The 'tripos' speeches for 1686 and 1687, both preserved in manuscript in the British Library, seem to be, in the same way, the work of several hands. The intimacy of a student coterie is intensified by the fact that the members are all taught by the same lecturers so that they share the knowledge they have gained in the classroom as well as familiarity with the oddities of the objects of the satire – the lecturers. Parody plays a large part in these texts.

However, the most virulent and subversive verse satires of seventeenth-century Ireland were probably produced outside Trinity College and, indeed, outside Dublin. Between 1665 and 1689, three versions of a ferociously anti-Irish and anti-Catholic satire were circulating in a closed circle of New English Protestants based in Fingal, in the Pale north of Dublin. This is a burlesque verse travesty of the sixth book of the *Æneid* – the book in which Æneas visits the underworld and meets his father. The poem occurs in three versions, one dating from about 1665, one from about 1686 and the other from 1689. Its earliest version (now National Library of Ireland MS 470, previously Heber 1568 and Phillips 8366), contains references to the enclosing of St Stephen's Green in Dublin – which took place in 1665 – so the text can not be earlier than that; it is entitled 'Purgatorium Hibernicum, or, The sixt book of Virgill's Æneis travestie burlesque a la mode de Fingaule': this version of the text consists of about 3,000 lines of rhyming tetrameter. The second, shorter version of the text (now British Library, Sloane MS 900, and again in rhyming lines of four stresses), is adapted to the early part of the reign

45 Carpenter, '*A Tale of a Tub* as an Irish text' forthcoming in *Swift Studies*.

of James II; it contains references to Tyrconnell and his imminent take-over of the Irish army and is entitled 'The Fingallian travesty, or, The sixt book of Virgills Æneid a la mode de Fingaule'. The third version, printed in London in 1689 and considerably shorter than either of the other two versions, is entitled *The Irish hudibras, or, Fingallian prince, taken from the sixth book of Virgil's Ænæds, and adapted to the present times.* The story in all three versions is of Nees (cf. Æneas), a prince of Fingal who descends to the underworld (St Patrick's Purgatory) through a cleft in the rocks off St Patrick's Island near Skerries (a fishing-port in Fingal), accompanied by Dydy (cf. Dido), a coarse ex-nun; after sundry adventures (closely paralleling those in *Æneid* VI) in Purgatory, Nees finds his father who prophesies the end of the Irish race. Nees then emerges from the underworld and the poem ends in various ways. The liveliest ending is in the *Irish hudibras* where, while Nees' forces plunder three Dublin pubs – the Mermaid, the Unicorn and the Saracen's Head – he himself overnights grandly in the 'Cabash' or Globe Inn. The objects of the satire in all versions of the text are the Old English who have become Hibernicized, their way of speaking English (both regular Hiberno-English and the exaggerated speech characterized in the text as 'Fingallian'), their lifestyle, their pretentiousness (Nees himself is titled a 'prince') and their Roman Catholicism. Throughout the text, Nees and his companions (including Macshane his piper (cf. Misenus) who is killed for farting) are portrayed as anti-heroes, comic, vain and gullible.

The most interesting aspect of these texts, particularly of the 'Purgatorium Hibernicum', is that they demonstrate a high level of cultural and linguistic interaction between English-speaking and Irish-speaking inhabitants of Restoration Ireland – an interaction not reflected to any significant extent in any other poetic text produced in seventeenth-century Ireland. The poems themselves are written in Hiberno-English, but this is interspersed with long, almost unintelligible passages of high-flown 'Fingallian' speech which spurt from Nees' mouth at regular intervals. The author (or 'authors') who have strong sympathies with the New English Protestants, were intimately acquainted with the Hiberno-English spoken in seventeenth-century Fingal and with the way of life there, of which hundreds of examples are given. The text is highly ingenious, highly energetic and full of the kind of vulgarity which finds it funny to rhyme 'sturdy' with 'a turd she'; its length makes the poem as a whole rather tedious but, in small doses, it can be entertaining. The earliest version, the 'Purgatorium Hibernicum', contains many Anglicizations of Irish words unrecorded elsewhere. It is remarkable that the texts contain puns only intelligible to readers who know both English and Irish, and in places, the English text is glossed into a phonetically rendered Irish; at line 134 of the 'Fingallian Travesty', for instance, 'To water head with yellow young' is glossed in the margin in phonetic Irish *Chon uska Pye oge*. To get the joke, you must translate each separate word from English to Irish and join *buí* and *óg* ('yellow' and 'young') to make *buíóg* ('bucket'). These examples of marginal 'clarification' for Irish-speakers of bilingual puns made in English raise fascinating questions about the intended readership for this text – which must have been trilingual. Readers were expected to have enough Irish to understand the English-Irish jokes and be familiar with the Irish-language manuscript histories of Ireland referred to in the notes (histories by O'Sullevan, Mulrony and O Mulconry), enough English to read accounts of Ireland by Spenser and Holinshed and understand the text itself and enough Latin to make sense of long quotations from historians of Ireland writing in that language, O'Sullivan Beare in particular. Their Latin must also have been good enough for them to enjoy the running joke of the relationship between the Virgilian Latin at the foot of each page and the outrageous Hiberno-English tetrameters above it. The texts differ from each other in extraordinary ways and were updated at least three times in their twenty-four year history; this suggests that several authors reworked the texts and that they circulated in a closed group for at least a generation. It is at present unclear what was the intended readership for these texts – particularly the two earlier manuscript versions – and who the authors were: Irish historians such as Sir James Ware and Archbishop Ussher had access to Irish historical material in all three languages earlier in the century, but it is hard to envisage the membership of a coterie in which multilingual historical knowledge and a passion for travesty would co-exist in Restoration Ireland over such a long period. Lawyers returned from the London Inns of Court or senior students at Trinity College, Dublin would have enjoyed and understood the jokes; but work remains to be done on the identification of individuals born in Fingal who might have been active in literary or legal circles in Dublin at the time. When we find these individuals, we may be able to reconstruct the coterie in which the various versions of this poem circulated; we could then find out who wrote this travesty, who were its intended readers and what exactly it meant, from a political, social and sectarian point of view.

Finally, it is worth pointing out that, as the seventeenth century came to a close, Jonathan Swift was writing verse satires as a member of the community in Dublin Castle. Several of his best early satirical poems come from this period, particularly 'Mrs Harris's petition' – a remarkable poem which was clearly written for circulation in the household of the Castle. The text bursts with specific references and personal details. Swift scholars have noted that these early poems of his circulated in manuscript for several years before they were caught by the printing press and became part of the Swift 'canon'. There were other, anonymous, poets active in Dublin Castle in the 1690s too – one of the poems, 'The picture of a Beau', tailored specifically to the personnel in Dublin Castle in the period 1699–1700, circulated in manuscript until the twentieth century – and I suspect that, if one were to search for them, one could find copies of other poems from Dublin Castle dating from the sixteenth and seventeenth centuries. There is a lot of work to be done on the Irish communities in which verse in English was written and in which it circulated – the coteries, we should call them; the fact that they existed so actively in Dublin Castle, and indeed in many other parts of the country, needs to be incorporated into any account of the culture of early modern Ireland.

Gathering ideas

A clerical library in County Cork, 1774

TOBY BARNARD

THE HINGSTONS WERE A FAMILY of a type common in Ireland between the seventeenth and early twentieth centuries. James Hingston, English in origin, arrived in Ireland with the Cromwellian army of reconquest late in the 1640s. Irish Catholics were being expropriated, so the seventeenth century was a propitious moment for newcomers to acquire lands. In the aftermath of the Williamite War, with the confiscation of Lord Clancarty's lands in County Cork, William Hingston duly purchased an estate at Aglis near Coachford. It consisted of 353 acres. In a second strategy used frequently by those of like background, successive generations of Hingstons married into other Protestant families of the locality: the Morleys, an aldermanic dynasty in Cork city; and the Webbs of Aglis. This practice continued into the eighteenth century when the Hingstons connected themselves by marriage with grander landed families in south Munster, such as the Percevals, Hulls, Bechers, Bernards and Claytons. A third device common to Protestant settlers was to make careers in the established Church of Ireland. This strategy was also adopted by the Hingstons. The Reverend James Hingston, having been ordained in 1736, served as

I am grateful to Andrew Carpenter, David Fleming, Raymond Gillespie, Marie-Louise Legg, Rolf Loeber and Raymond Refaussé for help on various points. English short-title catalogue [ESTC] numbers are given when relevant.

vicar of Clonmeene from 1751 until he died in 1775; his son, also a James, worked as diocesan schoolmaster in Cloyne, was appointed vicar-general of the diocese in 1794 and instituted as rector of Aghaballoge in 1799; a third James Hingston was also ordained deacon in 1806. The tradition survived into the twentieth century when yet another James rose to be archdeacon of Cork between 1948 and 1954.[1]

The first clerical James Hingston attracted little attention outside his locality. Given this obscurity, one detail of his life assumes greater interest. Hingston owned a substantial library, which was catalogued systematically in 1774.[2] The need for the clergy in their professional capacities to buy and read books is well known. They, together with other professionals such as medical doctors and lawyers, formed the largest identifiable groups to own substantial libraries in eighteenth-century Ireland.[3] In general, the bookish divines who have been studied were either bishops – Samuel Foley, William King, Narcissus Marsh, Thomas Percy, John Stearne and Edward Synge – or clerical scholars intent on research and writing.[4] Hingston is unusual in that he belonged to neither of the groups of the clergy usually associated with the making of large libraries. Yet, by 1774, he had a collection of approximately 2,500 titles. This did not approach the 7,000 assembled by Archbishop King of Dublin before his death in 1729 or by Edward

1 W.M. Brady, *Clerical records of Cork, Cloyne and Ross*, 3 vols (Dublin, 1863–4), ii, pp. 3–4, 188–9, 406–7; G.D. Burtchaell and T.U. Sadleir, *Alumni Dublinenses* (Dublin, 1935), p. 401; H. Cotton, *Fasti Ecclesiae Hibernicae* 6 vols (Dublin, 1845–78), i, pp. 294, 298, 306, 309–10, 312, 316.

2 Representative Church Body Library [hereafter RCBL], MS 521.3.

3 *The life and errors of John Dunton*, 2 vols (London, 1818), ii, pp. 496, 516–25; R.C. Cole, 'Private libraries in eighteenth-century Ireland', *Library Quarterly*, 44 (1974), 231–47; M. Pollard, *Dublin's trade in books, 1550–1800* (Oxford, 1989), pp. 214–16

4 On this group, see T. Barnard, 'Scholars and antiquarians: the clergy and learning, 1600–2000' in T. Barnard and W.G. Neely (eds), *The clergy of the Church of Ireland, 1000–2000* (Dublin, 2006), pp. 231–58.

5 *Catalogue of the Cashel diocesan library* (Boston, 1973); R.S. Matteson, 'Archbishop William King and the conception of his library', *The Library*, 6th series, 13 (1991), pp. 238–53; R.S. Matteson, *A large private park. The collection of Archbishop William King 1650–1729*, Libri Pertinentes, no. 7, 2 vols (Cambridge, 2003). For information about Synge's books I am deeply indebted to Dr Marie-Louise Legg who is engaged on a detailed analysis of them.

Synge, archbishop of Tuam from 1716 to 1741, who owned at least 2,635 titles.[5] However, Hingston's tally comfortably exceeded the 1,800 owned by King's contemporary and friend, Samuel Foley, briefly bishop of Down and Connor, and easily surpassed libraries of other early eighteenth-century bishops. Simon Digby, bishop of Elphin until 1720, had a collection of about 600 in his two Connacht residences.[6] A successor of Digby, Robert Howard, left in his 'study' at Elphin more than 380 volumes.[7] Another bishop of Down and Connor, Francis Hutchinson, gathered a collection which, when auctioned in Dublin in 1756, almost twenty years after Hutchinson had died, numbered about 600 titles.[8] Nearest to Hingston's library, both in size and the status of its owner, was that of County Tyrone clergyman, James Hamilton. In 1781, this incumbent of Donagheady was said to own about 2,000 volumes.[9]

A summary analysis of the titles recorded by Hingston indicates a diversified collection. The very diversity poses numerous puzzles: sources of supply, dates and modes of acquisition, and (above all) the uses of and responses to the volumes. Books arrived by a variety of routes: inheritance and gifts as well as more purposeful purchasing. Many titles in Hingston's collection were already a century (or more) old. It is possible that the books had been acquired by forbears when new or bought later when second-hand. Seventeenth-century Ireland offered opportunities to put together impressive libraries, thanks particularly to the enforced dispersal of collections made by Catholic priests and religious houses. By the 1690s, a growing demand, especially

6 Inventories of Bp S. Digby's goods at Abbert and Lackan, *c.*4 July 1720, National Library of Ireland [hereafer NLI], French of Monivea MSS, envelope 26.

7 Inventory of Bp R. Howard's books, 12 June 1740, NLI, PC 223 (6); receipt from Patience Howard to Bp E. Synge, 20 Feb. 1740[1], ibid., PC 225 (4); M.L. Legg (ed.), *The Synge letters: Bishop Edward Synge to his daughter, Alicia, Roscommon to Dublin, 1746–1752* (Dublin, 1996), p. xx.

8 *A catalogue of the books: being the library of the Right Rev. Dr. Francis Hutchison, late bishop of Down and Connor. To be sold by auction, by William Ross at the coffee-house of … the House of Lords* (Dublin, 1756), annotated copy in Queen's University, Belfast, WP/66 Z 997/1; G. Wheeler, 'Bishop Francis Hutchinson: his Irish publications and his library' in J. Gray and W. McCann (eds), *An uncommon bookman: essays in memory of J.R.R. Adams* (Belfast, 1996), pp. 140–55.

9 Public Record Office of Northern Ireland, D 671/D8/5/5 (I owe this reference to William Roulston).

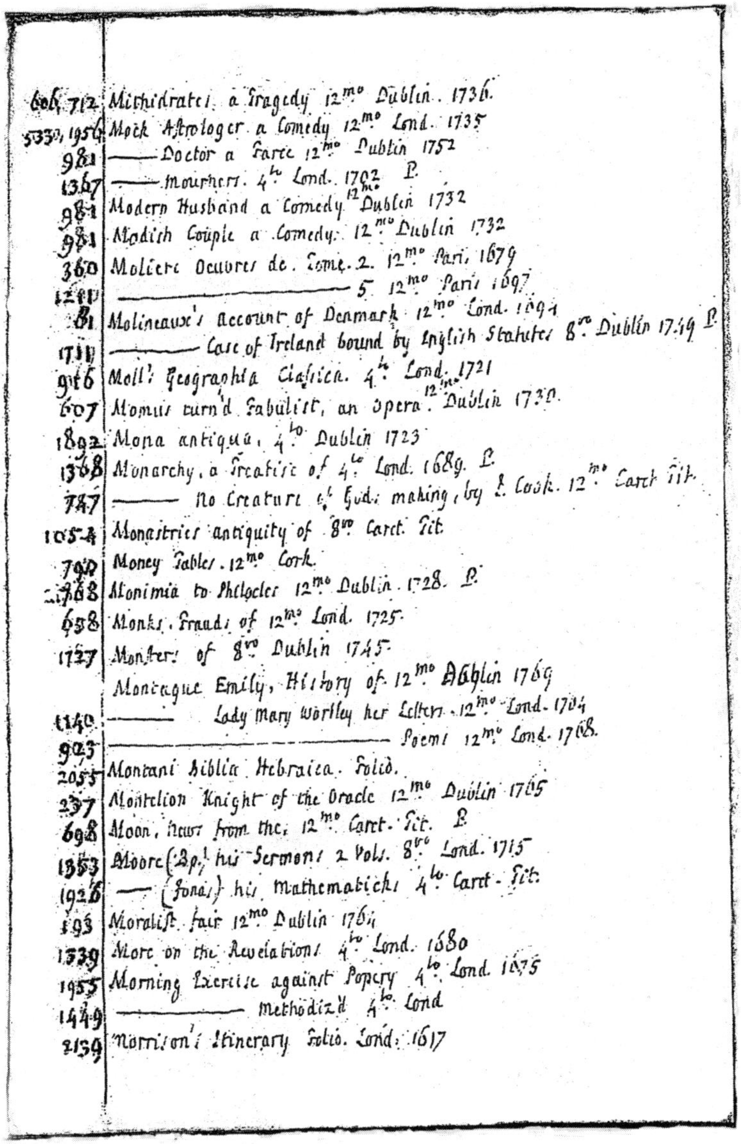

606, 712 Mithridates, a Tragedy 12mo Dublin. 1736.
5330, 1956 Mock Astrologer a Comedy 12mo Lond. 1735
981 —— Doctor a Farce 12mo Dublin 1752
1367 —— mourners. 4to Lond. 1702 P.
981 Modern Husband a Comedy 12mo Dublin 1732
981 Modish Couple a Comedy. 12mo Dublin 1732
360 Moliere Oeuvres de. Tome. 2. 12mo Paris 1679
1211 —————— 5. 12mo Paris 1697
81 Molineaux's Account of Denmark 12mo Lond. 1694
1711 —— Case of Ireland bound by English Statutes 8vo Dublin 1749 P.
916 Moll's Geographia Classica. 4to Lond. 1721
607 Momus turn'd Fabulist, an opera 12mo Dublin 1730.
1892 Mona antiqua, 4to Dublin 1723
1368 Monarchy, a Treatise of 4to Lond. 1689. P.
787 —— No Creature of God's making, by E. Cook. 12mo Caret Tit.
1054 Monastries antiquity of 8vo Caret Tit.
790 Money Tables. 12mo Cork.
1788 Monimia to Philocles 12mo Dublin. 1728. P.
658 Monks. Frauds of 12mo Lond. 1725.
1727 Monsters of 8vo Dublin 1745.
Montague Emily, History of 12mo Dublin 1769
1140 —— Lady Mary Wortley her Letters 12mo Lond. 1704
923 —————— Poems 12mo Lond. 1768.
2055 Montani Biblia Hebraica. Folio.
237 Montelion Knight of the Oracle 12mo Dublin 1705
698 Moon, news from the, 12mo Caret. Tit. P.
1353 Moore (Bp.) his Sermons 2 Vols. 8vo Lond. 1715
1926 —— (Jonas) his Mathematicks 4to Caret. Tit.
193 Moralist. fair 12mo Dublin 1764
1339 More on the Revelations 4to Lond. 1680
1955 Morning Exercise against Popery 4to Lond. 1675
1449 —————— methodiz'd 4to Lond
2159 Morrison's Itinerary Folio. Lond. 1617

1 Extract from James Hingston's library list, 1774,
courtesy of the Representative Church Body Library

among Protestant clerics, stimulated public auctions of books, chiefly in Dublin. In some cases, what was printed locally was sold; but stock was also imported specially for the sales. The professional preoccupations of a clergyman are reflected in the collection, but so too are those of the women and children of the household. The library included, for example, two copies of *The compleat housewife* – hardly the staple of a grave minister of religion.[10] In addition, as will be stressed, the library was rich in printed novels and plays. Serendipity, opportunism, gifts, even plunder, may explain some of the contents; so, too, do the accumulations and changing tastes of several generations.

The making of the catalogue is an act that deserves comment. It is methodical, and achieves a high level of bibliographical accuracy. One motive may have been to make the library more accessible by enabling particular volumes to be located more quickly. Another reason for cataloguing the collection could be that intimations of mortality were besetting Hingston, who would die during the next year. The list might be the prelude to dividing the books between his heirs, two of whom were also destined for the ministry. However, a different possibility is raised by the format of the catalogue. In the main, entries are written on the left-hand page. But, a few titles have been added on the opposite, otherwise blank side. One such addition is dated 1776, two years after the date on the title-page of the manuscript catalogue. It is in the same hand as the other entries, and so suggests that the inventory was compiled by the younger James Hingston, either just before, or at the moment when, the collection was handed to him by his father.[11]

The catalogue notes places of imprints and the size of the volumes. The separate titles of items bound with others are recorded. An alphabetical notation was added after some entries: 'B', 'C', 'D', 'E', 'H', 'L', 'M', 'P'. The signification of most letters is obvious: 'B' means biographies; 'C' for Greek and Roman classics; 'D' denoted divinity; 'H' for histories; 'L' for law. Less clear is his use of 'E', which may have been applied to literary works in English, and 'P', which seems to have been

10 Presumably the work by Eliza Smith.

11 'Gratian' [Baltasar Gracián], *The complete gentleman* (n.p. [Dublin], 1776) [ESTC T109129].

used for both pamphlets and poetry. 'Antiq', 'Math', 'Ph' and 'Pol' are clear enough, indicating, respectively, antiquities, mathematics, physic (or medicine) and politics. Because Hingston's eagerness to classify the entries in this manner quickly faded, only the opening pages of the catalogue use the annotations systematically. His waning interest makes it difficult to calculate the proportions of books in each category.

II

Most predictable among the components of the collection was what catered directly to the Hingstons' clerical calling: theology. Works by English apologists of the established church, such as Lancelot Andrewes, William Beveridge, Gilbert Burnet, William Chillingworth, Henry Hammond, George Herbert, Simon Patrick, William Sherlock, Robert Saunderson, Thomas Secker, Edward Stillingfleet, Jeremy Taylor and John Tillotson abounded. Books like Burnet's *Pastoral care*, reprinted in Dublin in 1734, or *The parson's Christian directory* and Bull's *A companion for candidates for holy orders, or, The great importance and principal duties of the priestly office*, both English publications, had obvious practical applications. The need on occasion to defend the distinctive position of the Church of England and its sister establishment in Ireland made useful the likes of Revd Thomas Breviter, *No democracy in the church*[12] and *A preservative against popery*[13] or William Nelson, 'Rights of the clergy'.[14] The same controversial tenor ran through Hales' 'Best way of refuting papists'.[15] Locally, alarm was aroused by the seemingly heretical opinions expressed in a tract of 1752 by Robert Clayton, formerly bishop of Cork but since translated to Clogher. Hingston possessed four refutations of Clayton's *Essay on spirit* as well as the offending original.[16]

12 Dublin, 1735 [ESTC T88774]. This is a rare example of Hingston making an error in his catalogue. He transposed the publication dates of the two Breviter tracts.

13 Dublin, 1728.

14 London, 1709.

15 London, 1708.

16 R. Clayton, *An essay on spirit* (Dublin, 1752); W. Jones, *A full answer to the essay on spirit* (Dublin, 1755); Charles Leslie, *Deism refuted* (Dublin, 1755); T. McDonnell, *An essay towards an answer to a book, entitled, an essay on spirit* (Dublin, 1753); and an anonymous answer published in Dublin in 1752.

In addition to classics of Anglican devotion and apologetic, there was a substantial store of the early Christian fathers and of recent Catholic controversialists, like Bellarmine. The first category included St Clement's *Epistolae ad Corinthius*,[17] the works of St Cyprian, a folio printed at Paris in 1573, a more recent edition of Cyprian 'on the unity of the church',[18] St Isidore's works in a folio printed at Madrid in 1699 and the staple of Eusebius' *Ecclesiastical History*.[19] Hingston owned four copies of Bellarmine, three of which had been published in Amsterdam between 1636 and 1648, and two different editions of Aquinas, *Summa Theologiae*: a folio, printed at Douai in 1614; and a three-volume version published in Paris in 1657. He had Father Paul's *History of the Council of Trent* and the Louvain Testament, bearing the imprint of Rouen, 1663. In the catalogue, under the heading 'popery' twenty-three volumes were detailed. Again the puzzles as to who first brought these venerable tomes published in continental Europe into Ireland and when they were added to Hingston's holdings are, at least for the moment, unanswerable riddles.

To protect the other flank, Hingston had ten titles appertaining specifically to the Protestant dissenters and their alleged errors. These non-conformists, far less numerous than Catholics in Hingston's parishes, nevertheless constituted a danger, whether cosmically, nationally or locally. Further insights into local competitors were afforded by items like Barclay's *Apology for the Quakers*, the Quaker Isaac Pennington's *The light within* and Lodowick Muggleton's *Divine looking glass*. The Muggletonians, developing from the Quakers in the late 1650s, had enjoyed a brief vogue in County Cork, so it may not have been hard for one of the Hingstons to pick up a copy of the tract.[20] The Quakers were formidable enough in the area for Hingston to equip

17 Oxford, 1677.

18 Paris, 1573; Oxford, 1681.

19 Cf. B. Neveu, 'L'érudition ecclésiastique de xviie siècle et la nostalgie de l'antiquité Chrétienne' in K. Robbins (ed.), *Religion and humanism*, Studies in Church History, 17 (1981), pp. 195–225; J-L. Quantin, 'The Fathers in seventeenth-century Anglican theology' in I. Backus (ed.), *The reception of the Church Fathers in the West* 2 vols (Leiden, 1997), ii, pp. 996–9.

20 P. Kilroy, *Protestant dissent and controversy in Ireland, 1660–1714* (Cork, 1994), pp. 87–8.

himself with at least ten works about them. The easiest explanation for the presence of these volumes in the library is the persistence of confessional rivalry and the need to know the teachings of adversaries if they were to be routed. Yet, it is not the sole explanation. A less partisan curiosity about divergent beliefs may have motivated Hingston, as other Irish Protestants with theologically eclectic collections.

Many other volumes related to the sacred ministry. For example, Hingston recorded four separate expositions of the Thirty-Nine Articles, the bed rock of Church of England and Church of Ireland belief and practice. He had versions of the Bible in Greek,[21] Hebrew, two in Latin,[22] the Vulgate, and a folio edition in English, printed in Dublin in 1714. He also had eleven versions of the New Testament, including the Irish translation of 1681 and a Dublin printing of 1746.[23] To these were added two concordances, two histories of the sacred text and The Westminster Assembly's annotations on it.[24] There were Common Prayer Books in Greek and Latin, the Irish edition of 1692 and four editions of the English version, as well as six commentaries on the Prayer Book; seven catechisms, of which two were printed in Dublin and two in Cork; visitation articles from England;[25] four printed charges from bishops (three of them occupants of Irish sees); and a 1748 set of instructions for candidates for confirmation. He had a copy of Calvin's *Institutes*, published at Geneva in 1606: a possible sign that some in the Church of Ireland still leant towards Calvinism.

Sermons were a genre usually well represented in clerical – as indeed in many lay – libraries. Hingston's was no exception. He had four volumes of those preached on the benefaction of Robert Boyle and another collection of the philippics against popery delivered in Salters' Hall in London.[26] Then there were four quarto volumes published in Dublin and another two in octavo format, also issued in Dublin. Four other collections of printed sermons by unnamed divines were noted.

21 2 vols, Cambridge, 1665.

22 London, 1611, and in four volumes, London, 1726.

23 He also had William Fulke's critique of the Rheims New Testament. Fulk, *Rhemish Testament*, 4°, no title page.

24 He also owned the Assembly's catechism and 'Directory' [of worship].

25 London, 1726.

26 Dublin, 1737; Dublin, 1735.

He also owned works by many of the most famous Church of England preachers of the seventeenth and eighteenth centuries, such as William Beveridge, Samuel Walker and Seth Ward.[27] He could muster eight volumes of Tillotson's sermons.[28] Instruction and edification were to be derived by the minute perusal of sermons. For other clerics, such compendious publications served as models for their own compositions. Since none of the ecclesiastical Hingstons is known to have published a sermon, the style that they favoured cannot now be ascertained.

Local concerns could hardly be separated from Hingston's sacerdotal duties. Conscientious incumbents endeavoured to educate and relieve parishioners. The charter school movement was one initiative intended to help the poor and ignorant. Hingston was interested enough to have the 1713 account of its progress, a 1726 guide, the 1734 report of the proceedings in relation to the charity, and a 1739 collection of the sermons delivered annually to raise funds.[29] There were also two printed reports of the charities established to support widows of the clergy in, respectively, the dioceses of Dublin and Cloyne (his own). Interest in the affairs of his diocese was further suggested by three of the manuscripts that he recorded: the acts of the dean and chapter of Cloyne; the state of the charitable fund for the diocese; and a succession list of its clergy. With the first document, it is unclear whether he had removed or 'borrowed' an original chapter act book or merely made his own extracts. The presence of the last indicated antiquarian interests which may go some way to explain the size and variety of his library. Hingston's unpublished account of Cloyne in 1762 was used extensively by W.M. Brady in the mid-nineteenth century when the latter constructed a biographical register of the dioceses of Cork, Cloyne and Ross.[30]

27 2 vols, London, 1720.

28 3 vols, London, 1714, and 5 vols, London, 1720.

29 See K. Milne, *The Irish Charter Schools, 1730–1830* (Dublin, 1997).

30 Brady, *Clerical records*, ii, p. 189. Hingston's manuscript is not among the surviving archive of the diocese of Cloyne. RCBL, C.12; D.16; S. Hood, 'The diocesan records of Cloyne, County Cork: an unlikely source of genealogical significance in the Representative Church Body Library, Dublin', *Irish Genealogist*, 9 (1994–7), 426-37. Only the return for 1791 by his son, James Hingston, as rural dean of Cloyne now attests to the investigations of the family. RCBL, D.16, 10/1.

Education of a different kind is also reflected strongly in the library. By the eighteenth century, the clergy of the Church of Ireland were overwhelmingly university graduates, most commonly of Trinity College, Dublin. Hingston had been graduated at Trinity in 1734, and two of his sons followed him thither.[31] Accordingly, they were equipped with the standard learning necessary both to enter and graduate from the college. Greek and Roman classics bulked large in this preparation and in Hingston's collection.[32] The cataloguer noted a five-volume Greek collection of Aristotle's writings printed at Francovia in 1537; a Greek and Latin edition, also in five volumes, in octavo, published at Hanover in 1611; and a Paris printing of the *Ethics* of 1558. Additionally, and possible relics from student days, there were Aristotle's *Problems*[33] and a folio manuscript of his *Historia animalium* translated into English. Another manuscript was of notes on 'college classicks'. Other texts on logic and rhetoric and the numerous editions of Cicero, Euclid, Homer, Horace, Juvenal, Ovid and Virgil, may have been relics of university study, retained for the use of subsequent generations. All six treatises on rhetoric found in the library had been published in the seventeenth century: another indication of a brisk trade in used textbooks. The demands of an orthodox pedagogic regime are suggested powerfully by the Hingstons' ownership of handwritten versions of the *Iliad*, *Odyssey*, Juvenal's and Lucian's *Works*. They were probably either exercises or cribs.

The sheer size and variety of Hingston's library far excelled the collections routinely assembled for and by intending ordinands. What may have started as the necessities for study at university expanded into a much larger library. The many dictionaries and grammars exemplify the bibliomania. Clerics were expected to read in Greek, Latin and perhaps Hebrew, as well as in the modern European languages of French

31 Burtchaell and Sadleir, *Alumni Dublinenses*, p. 401.

32 V. Kinane, *A history of the Dublin University Press, 1734–1976* (Dublin, 1994), pp. 64–5; J.W. Phillips, *Printing and bookselling in Dublin, 1670–1800* (Dublin, 1998), pp. 290–4; M.J. Ryan, *List of Greek and Latin classics printed in Dublin*, Bibliographical Society of Ireland (Wexford, 1926), W.B. Stanford, *Ireland and the classical tradition* (Dublin, 1984), pp. 49–51.

33 London, 1700.

and Italian. Hingston possessed volumes in all these languages: he itemized three French, five Greek, six Hebrew, seven Latin and one English grammar. Perhaps most tantalizing in its implications is his ownership of two English-Irish dictionaries: that printed in Paris in 1732, and O'Brien's, also printed in Paris (in 1768). A recent purchase was Vallancey's Irish grammar that had been published in 1773, only one year before the catalogue was compiled.

III

Hingston's interest in Ireland also went beyond the conventional concerns of a parochial minister labouring in a difficult environment. Involvement in south Munster no doubt accounted for his ownership of the three county histories of Cork, Kerry and Waterford by the Dungarvan apothecary, Charles Smith. Hingston had subscribed for the Cork volume, published in 1750. In addition, and widening his horizons, he had the companion volumes on Counties Down and Dublin, the later history of Limerick by John Ferrar and Sir William Petty's maps of Ireland in *Hiberniae delineatio*. Locality was represented too with a pamphlet urging the improvement of the navigation on the River Blackwater.[34] He had copies of one of the younger Sir Richard Cox's published charges to the County Cork grand jury and of his own bishop, George Berkeley's, exhortation to the county magistrates. He also owned effusions by 'Alexander the Coppersmith' (the pseudonym of William Boles) and a later imitator on the contested corporation and parliamentary elections for Cork city in 1731 and 1751.[35] Another dispute in Kinsale, as the agitation over the Money bill in 1753 spread from the

34 Richard Brocklesby, *A letter to a member of the Irish Parliament ... extending the navigation of the Blackwater river* (Cork, 1755). Bound with it was *The utility of inland navigation* (Cork, 1755).

35 'Alexander the Coppersmith' [W. Boles], *Milk for babes, meat for strong men and wine for petitioners. Being a comical, sarcastical, theological account of a late election at Bagdad for a bailiff of the city* [Cork, 1731]; *A ramble through Bagdad in a letter from Philologos to his friend* (Bagdad, i.e. Cork, 1751); E. R. McC. Dix, 'List of books, etc. printed in Cork in the 17th and 18th centuries', *Journal of the Cork Historical and Archaeological Society* [hereafter *JCHAS*], 6 (1900), 236.

Dublin parliament into the provinces, occasioned one pamphlet owned by Hingston.[36]

Otherwise, to judge from the collection, contemporary political controversies were not a prime concern. Hingston kept a few partisan publications from Queen Anne's reign, when many churchmen had been active combatants. There was another brace of pamphlets relating to the controversy over the patent to mint copper coin for Ireland granted to William Wood in the 1720s.[37] In the late 1740s, Charles Lucas and Sir Richard Cox used print to pursue public and private objectives, but Hingston had few of the effusions notwithstanding Cox's strong Cork ties.[38] A scheme of 1751 for a legislative union of Great Britain and Ireland did find its way into the library.[39] Also topical may have been a history of the Cavendish family published in 1756. A new lord lieutenant, Lord Hartington, eventually duke of Devonshire, who arrived in 1755, headed the Cavendishes.[40] In addition, Hartington was linked to the locality as inheritor of the vast Boyle holdings in south Munster. The combination may account for Hingston's interest. Within Ireland, the Boyle affinity was controlled by Henry Boyle, as Speaker of the Commons and leader of a squadron of MPs many of whom hailed from Munster. Hingston owned one publi-

36 T.C. Barnard, 'Considering the inconsiderable: electors, patrons and Irish elections, 1659–1761' in D.W. Hayton (ed.), *The Irish parliament in the eighteenth century: the long apprenticeship* (Edinburgh, 2001), pp. 107–27; T.C. Barnard, 'The cultures of eighteenth-century Irish towns' in P. Borsay and L. Proudfoot (eds), *Provincial towns in early modern England and Ireland; change, convergence and divergence*, Proceedings of the British Academy, 108 (2002), pp. 215–21. Otherwise, almost the sole reminder was *Ministerial conduct of the governors of Ireland* (Dublin, 1754).

37 *Answer to High Church standers* (London, 1711); *High Church Men's honesty exemplified* (London, 1707); *Defence of William Wood* (Dublin, 1725); *Defence of the Irish for refusing the coin of William Wood* (Dublin, 1724).

38 C. Lucas, *Letters to the Commons of Dublin* (Dublin, 1749). There was also in the same volume, an address from James Digges La Touche, Lucas's partner in fighting the Dublin parliamentary election of 1749.

39 J. Kelly, 'The origins of the Act of Union: an examination of unionist opinion', *Irish Historical Studies*, 35 (1987), 236–63.

40 *Memoirs of the family of Cavendish* (1756). A Dublin edition of White Kennett's work of this name is known, but not of this date.

cation that reviewed Speaker Boyle's recent and controversial political role.[41] He had also acquired James Anderson, *History of the house of Yvery*, which detailed the tribe of Percevals, owners of substantial tracts of north Cork, into which Hingston had married.[42] Not so readily explained is a manuscript history of the house of Desmond, longer settled in Ireland than the Percevals or Hingstons. Local loyalties may again lie behind the acquisition of a six-volume set of Spenser's *Works*: the poet's Cork connections had recently been exploited in a bid to publish an Irish edition.[43]

More distant events left traces in the library. A publication of 1740 detailed Britain's mistakes in the conduct of war. Others investigated Admiral Vernon's carriage at Portobello, British and allied operations in Europe during 1743 and 1744, and Admiral Matthews' activities in 1744. English and French territories in North America were described in a Dublin production of 1754, on the eve of conflict there. More recent still were Admiral Byng's attempts to justify his conduct in a letter to the Admiralty and General Bligh's letter to William Pitt, printed in Cork during 1758. Another event that led to printed reflections – by John Wesley – which found their way onto Hingston's shelves was the Lisbon earthquake of 1756.[44] Curiosity about British as distinct from Irish politics can be detected in copies of Pulteney's letters to Sir Robert Walpole and of a key to 'an emblematical portrait' of Walpole.[45] Both topical and more local was a 'proposal for lessening the price of bread corn'. Published in Dublin in 1757, the pamphlet addressed the latest in a sequence of subsistence crises in Ireland.

James Hingston combined civil with religious duties. He was commissioned as a justice of the peace for County Cork in 1759.[46] The

41 Perhaps *An account of the life, character and parliamentary conduct of the Right Honourable Henry Boyle, esq.* (Dublin, n.d.).

42 2 vols, London, 1742.

43 T.C. Barnard, 'Edmund Spencer, Edmund Spenser and the problems of Irish Protestants in the eighteenth century', in T.C. Barnard, *Irish Protestant ascents and descents* (Dublin, 2003), pp. 290–305.

44 John Wesley, *Serious thoughts occasioned by the late earthquake at Lisbon* (Dublin, 1756).

45 London, 1730; *Pulteney's five letters to Walpole* (Dublin, 1731).

46 H.F. Berry, 'Justices of the peace for the County of Cork', *JCHAS*, 3 (1897), 106.

responsibilities weighed heavily enough for him to acquire seven of the standard guides for magistrates. Two by Bolton and Dalton dated from the seventeenth century.[47] Dalton was intended primarily for those serving in England; so, too, were the manuals by Burn, Dudley and Nelson, all present in his library.[48] Only Bullingbrooke and Dutton aimed their handbooks specifically at Irish circumstances.[49] In addition, Hingston had copies of Dutton's treatise on the office of sheriff, Bullingbrooke's digest of Irish ecclesiastical law,[50] and Jacob's manuals for conveyancing, keeping courts and chancery procedure.[51] That Hingston treated this aspect of his local activities with particular seriousness is revealed by the survival of an abridgement that he himself made of the Irish statutes, probably as a vade-mecum for his work in the parish and on the magistrates' bench.[52] A handier guide was *Every man his own lawyer*.[53] Tithes were another complex matter, with potential for litigation, in which he was necessarily involved. He would be helped by Selden's and Prideaux's histories,[54] Bishop Edward Synge's published affidavits of 1736,[55] a scheme for abolishing tithes printed in 1742, and 'McCawley's three letters on tithes' of 1758.[56] In addition, Coghill on the state of the Irish first fruits and tenths (1723) and the 1741 Irish *Valor ecclesiasticorum beneficiorum* assisted the Hingstons in unravelling the tangle of financial dues.

William Hingston of Aglis was commissioned in 1773; the Reverend James Hingston, then of Cloyne, was appointed in 1797. Ibid, pp. 108, 109.

47 R. Bolton, *Justice of the peace* (Dublin, 1683); E. Bullingbrooke, *Ecclesiastical law*, 2 vols (Dublin, 1770); M. Dalton, *Country justice* (London, 1635).

48 Burn, 3 vols (London, 1762); Dudley and Cunningham, 4 vols (London, 1769); Nelson (London, 1704).

49 ?M. Dutton, *Justice of the peace* (Dublin, 1744); E. Bullingbrooke, *Justice of the peace* (Dublin, 1764).

50 Bullingbrooke, *Ecclesiastical law*; M. Dutton, *Sheriff and coroner* (Dublin, 1727).

51 All were published in London and aimed at English readers.

52 RCBL, MS 521.1.

53 Dublin, 1755.

54 H. Prideaux, *The original and right of tithes* (London, 1736); J. Selden, *The history of tythes* (London, 1618).

55 E. Synge, *Two affidavits in relation to the demands of Tythe-Agistment in the diocese of Leighlin* (Dublin, 1736).

56 Dublin, 1758, probably by Alexander MacAulay.

A few texts hinted at the Hingstons' place among a privileged but uneasy Protestant minority, both in the kingdom of Ireland and within their own region. Some, but not all, of the obvious histories of Ireland are there: Carew, *Pacata Hibernia*;[57] the two volumes of the letters of the first earl of Strafford, lord deputy of Ireland during the 1630s; Temple's account of the 1641 uprising, but only in the recent Cork reissue of 1766, rather than in any earlier edition; Borlase's *Reduction of Ireland*;[58] Clarendon's account of the same period, published rather belatedly in 1720;[59] and the more recently published *State letters* of Orrery, a principal protagonist in the war of the 1640s in Munster.[60]

Little of this history was surprising. More so was the presence of works that hint at the political stance of the Cromwellian founder of the Hingston dynasty in Ireland. They included the republican, Edmund Ludlow's *Letter on the tyranny of Charles I* and John Toland's uncompromisingly titled publication of 1698 'Charles I: no saint or martyr'.[61] Published in the same year as the latter was a defence of the Long Parliament (which had sat in the 1640s) against Charles I. In tune with these works that resonated with the spirit of true Whiggery, if not of outright republicanism, was Robert Molesworth's *Account of Denmark*.[62] It sounded a tocsin over a feared loss of liberty, not just in Denmark, but in Britain and Ireland. Another possible survival from the earliest phase of the Hingstons' lives in Ireland was 'Cook against monarchy'. Hingston's copy lacked its title page, but this was probably *Monarchy no creature of God's making* by John Cook and published at Waterford in 1651. Cook was enticed to Ireland by Cromwell to serve as chief justice of Munster between 1649 and 1655.[63] Hingston, in choosing to keep these republican texts, did not disavow the forbears who

57 London, 1711.

58 London, 1675.

59 He also had a six-volume octavo edition of Clarendon's *History of the Great Rebellion in England* (Oxford, 1712).

60 2 vols (London, undated).

61 D.J., King *Charles I. No such saint, martyr or good Protestant as commonly reputed* (London, 1698); J. Champion, *Republican learning: John Toland and the crisis of Christian culure, 1696–1732* (Manchester, 2003), pp. 94–5.

62 London, 1694.

63 T.C. Barnard, *Cromwellian Ireland* (Oxford, 1975), pp. 262–76.

had procured the works. At the same time, his attribution of Molesworth's *Account of Denmark* to Molyneux hardly betokened familiarity with its contents or ideological context.[64]

Another contentious matter when the Hingstons were establishing themselves in Ireland was the settlement of lands on the Cromwellian soldiery and settlers: a redistribution largely confirmed by the restored Charles II. The deal was impugned by a Catholic bishop, Nicholas French, in his *Narrative of the earl of Clarendon's settlement and sale of Ireland.* French's tirade, published at Louvain in 1668, was quickly banned by the Irish government. Hingston possessed a manuscript copy, maybe because his forbears had been unusually eager to read the controversial work.[65] He had a printed copy of French's other defence of Catholic Ireland in the mid-seventeenth century: *The bleeding Iphigenia,* but it lacked its title page.[66] The library contained two later Catholic histories – those of Reilly and the Abbé MacGeoghegan.[67] Less embattled antiquarianism is represented by a manuscript abridgement of Sir Robert Cotton's records in the Tower of London and a second manuscript described simply as 'historical collections'.

The Hingstons weathered the war between 1688 and 1691. It was described by Archbishop William King, George Story and Walter Harris, whose accounts were on James Hingston's shelves.[68] More local was a pamphlet proposing to send back to Ireland those nobility and gentry who had fled to Britain during 1688 and 1689. It provoked a

64 This rare mistake is repeated twice in the catalogue.

65 T.C. Barnard, '"Parlour entertainment for an evening": histories of the 1640s', in M. Ó Siochrú (ed.), *Kingdoms in crisis: Ireland in the 1640s* (Dublin, 2001), pp. 30–2.

66 Louvain, 1674.

67 N. Ó Ciosáin, *Print and popular culture in Ireland, 1750–1830* (Basingstoke, 1997), pp. 102–6; V. Geoghegan, 'A Jacobite history: the Abbé MacGeoghegan's *History of Ireland*', *Eighteenth-Century Ireland*, 6 (1991), 37–55; J.R. Hill, 'Popery and Protestantism, civil and religious liberty: the disputed lessons of Irish history, 1690–1812', *Past & Present*, 118 (1988), 96–129.

68 W. Harris, *A new history of the life and reign of William-Hennry*, 4 vols (Dublin, 1747); W. King, *The state of the Protestants under James II* (London, 1692); G.W. Story, *A true and impartial history of the most material occurrences in the kingdom of Ireland* (London, 1692).

riposte from a Cork landowner, St John Brodrick.[69] Yet, Hingston listed neither Brodrick's rejoinder nor a copy of the celebration of the subsequent Williamite triumph, *Hibernia Anglicana,* by the Cork lawyer and passionate Tory, the elder Sir Richard Cox. One repercussion from William of Orange's victories and his and his wife's ascent to the thrones of England, Ireland and Scotland was the question of abjuring their predecessor, James II, and swearing allegiance to the new sovereigns. The presence of five tracts reviewing the arguments in favour of recognizing William and Mary, published between 1689 and 1691, may show that the problem perplexed one of Hingston's ancestors. If this were the case, then, in common with most other Protestants in Ireland, the dilemma was resolved in favour of the new monarchs. Continuing enthusiasm for William III, at least in the posthumous cult, might be detected in the acquisition of Harris' eulogistic *Life* and of a pamphlet that urged the erection of a monument to him on the banks of the River Boyne.[70] A local controversy sparked off by Peter Browne while bishop of Cork, about toasting William III, was remembered through a copy of Browne's original criticism of the practice.[71] Absences from the collection included not only Cox's *Hibernia Anglicana*, but celebrations of the dukes of Ormond, the leading royalists of the later seventeenth century and eventually, in the case of the second duke, a Jacobite exile from Ireland.

Further local concerns are reflected in three manuscripts described as 'Mallow patent', 'Orange Club Book' and 'Blackwater Club'. In 1745, Hingston joined the 'loyalist' Protestant club established in Mallow, and served as its treasurer during the next year. Whether this body was the same as the 'Orange Club', a precursor, or a different organization is not clear.[72] The alarm that gripped the Protestants of south Munster

69 Sir R. Bulkeley, *The proposal to send back the nobility and gentry of Ireland* (London, 1690).

70 4 vols (Dublin, 1747); probably *A letter to … Sir Ralph Gore … concerning a lately published proposal for a voluntary subscription to erect a trophy in memory of the deliverance of this kingdom, by the glorious victory at the Boyne* (Dublin, 1732).

71 P. Browne, *Of drinking to the memory of the dead* (Dublin, 1715).

72 Minute book of Mallow Loyal Protestant Club, TCD, MS 7105, ff. 1v, 19v. This had previously belonged to the Bowen family, not to the Hingstons.

in 1745 echoed through the tract, *Seasonable advice to Protestants*, which tried to rouse them to greater vigilance. Hingston had his copy, as of a second publication designed to admonish insouciant Protestants, the 1766 Cork reprint of Sir John Temple's *History of the Irish rebellion*.[73] Timely, too, was the appearance in Dublin in 1745 of *Papal tyranny*. Hingston possessed four works aimed specifically against the Young Pretender, including *Honest advice against him to the Irish* and two printed accounts of the rebellion that he had inspired.[74] Participation in less overtly sectarian forms of associational life are implied by the ownership of a freemasons' *Handbook* and of *The fundamental laws of the Friendly Brothers of St Patrick*, the anti-duelling society that had gained popularity among the polite by the middle of the eighteenth century.[75]

IV

If public functions were catered for, so were private ones. The catalogue listed works for children: a geography; the *Child's new plaything* of 1774; two volumes of fairy stories, published in Dublin in 1750 and 1759; and an undated Dublin-printed history for children. Also suggestive of an interest in the proper upbringing of children are the four volumes of Jean-Jacques Rousseau's *La Nouvelle Héloïse*, which had appeared in Dublin in 1767.[76] It is possible that the clerical Hingstons taught oth-

73 Hingston subscribed to it, along with over fifty other Protestant clergymen from Munster. On the reprint, see T.P. Power, 'Publishing and sectarian tension in south Munster in the 1760s', *Eighteenth-Century Ireland*, 19 (2004), 75–110. For the longer history of the work: T.C. Barnard, 'Crises of identity among Irish Protestants, 1641–1685', *Past & Present*, 127 (1990), 39–83; T. Bartlett, 'A new history of Ireland', *Past & Present*, 116 (1987), 214–17; R. Gillespie, 'Temple's fate: reading *The Irish Rebellion* in late seventeenth-century Ireland', in C. Brady and J. Ohlmeyer (eds), *British interventions in early modern Ireland* (Cambridge, 2004), pp. 315–33.

74 Dublin, 1745.

75 *Freemasons Handbook* (Dublin, 1735); *Fundamental laws of the Friendly Brothers of St Patrick* (Dublin, 1751); Toby Barnard, *Making the grand figure* (London, 2004) pp. 361–6; J. Kelly, *'That damn'd thing called honour': duelling in Ireland, 1570–1860* (Cork, 1995), pp. 65–6.

76 On Rousseau's popularity: M. Kennedy, *French books in eighteenth-century Ireland* (Oxford, 2001), p. 139.

ers in the neighbourhood. One son, the Reverend William Hales Hingston, became diocesan schoolmaster in the diocese of Cloyne: a post for which his father's books had helped to prepare him and in which they would continue to assist. Works such as the 'manual for Winchester School', published in 1684, or Newton's *English academy* of 1677 would otherwise be surprising inclusions in the library. Arising as much from family as from parochial responsibilities were several medical books, which allowed the rudiments of care to be given within the household. These works included 'Cheyne on health',[77] Garth's 'Dispensary',[78] Nugent 'on hydrophobia', Richard Mead on small-pox and measles,[79] a Dublin-printed tract of 1744 on preventing deformities in children and writings by the Irish physician, Bernard O'Connor.[80] Hingston also had a manuscript of 'general maxims for health'.

For those who survived the dangerous years of infancy, instruction in polite conduct came from *The Spectator*, present in an eight-volume edition, and Lord Shaftesbury's *Characteristicks*.[81] Tutelage in more frivolous matters was offered by a Dublin edition of Edmund Hoyle's *A short treatise on the game of whist*, 'rules to aid the memory therein', printed in Dublin the following year,[82] and *The art of dancing*.[83] Anxiety over the last accomplishment is suggested, too, with the inclusion of the name of a Miss Hingston among the pupils of the leading dancing master in Cork, Laurence Delamain, in 1763. Delamain taught the young of respectable families from both the city and county: clearly the Hingstons fell into the second category.[84]

Clergy of the Church of Ireland had both practical and ideological

77 Dublin, 1725.

78 Dublin, 1725.

79 Dublin, 1748.

80 B. O'Connor, *Dissertationes Medico-Physicae* (Oxford, 1695).

81 L.E. Klein, 'Liberty, politeness and manners in early eighteenth-century England', *Historical Journal*, 32 (1989), 583–605; idem, *Shafesbury and the culture of politeness* (Cambridge, 1994).

82 Two editions of Hoyle were printed in Dublin in 1743: ESTC N22952 and ESTC N24789.

83 The date would suggest that this is the poem by Soame Jenyns printed in T. Odell, *The smugglers, a comedy* (Dublin, 1729) [ESTC T48470], rather than an unrecorded Dublin edition of Kellom Tomlinson's treatise.

84 Inventory of L. Delamain, 20 Jan. 1763, TCD, MS 2015/395.

reasons to interest themselves in the physical improvement of the country, on which, in the end, their own livelihoods as well as the celestial futures of themselves and their parishioners depended. The county histories by Harris and Smith, mentioned already, celebrated recent achievements. There are further hints that Hingston subscribed to the doctrine. He possessed Arthur Dobbs' *Essay on trade*, in which the patriotic creed was expounded.[85] He also owned Mandeville's *Fable of the bees*: the work which offered utilitarian reasons for the increasing consumption and production of luxuries. Other improvements might be practised thanks to John Evelyn's classic guide to the raising of trees, *Sylva and Pomona*,[86] the *Observations* of the Dublin Society, and associated manuals on gardening, tillage and the cultivation of fir trees, saffron, hops, hemp and flax.[87] In the same mode were two compositions by the Reverend Richard Barton, an enthusiast for improvements.[88] Another focus of improvers' ambitions, the nascent linen industry, was the subject of a propagandist pamphlet by the Cork landowner, Sir Richard Cox, and of a report by Robert Stephenson.[89] Yet, as in the case of the holdings on recent political controversies, Hingston seems to have selected sparingly from the plethora of titles on these subjects, hinting perhaps that he was not especially active in practising this fashionable form of patriotism.

In some instances, then, the collection is as noteworthy for what is sparsely represented, namely politics and improvements, as for what it

85 A. Dobbs, A*n essay on the trade and improvement of Ireland* (Dublin, 1729).

86 London, 1664.

87 Richard Bradley, *New improvements of planting and gardening*, 4th edn (Dublin, 1720); Richard Bradley, *The riches of a hop-garden explained* (Dublin, 1731); *Observations of the Dublin Society* (Dublin, 1739); *Instructions for raising hemp and flax* (Dublin, 1724); *Philosophical enquiry concerning hemp and flax* (Dublin, 1733); *The culture of saffron* (London, 1732); William Ellis, *The modern husbandman*, 3 vols (Dublin, 1743); ?W. Maple, *Tanning without bark* (Dublin, 1729). On this movement: P. Kelly, 'The politics of political economy in mid-eighteenth-century Ireland' in S.J. Connolly (ed.), *Political ideas in eighteenth-century Ireland* (Dublin, 2000), pp. 105–29.

88 R. Barton, *Description of Lough Lene* (Dublin, 1751); idem, *Lectures in natural philosophy* (Dublin, 1751).

89 *A letter from Sir Richard Cox, Bart. To Thomas Prior, Esq …* (Dublin, 1749); R. Stephenson, *The reports and observations* (Dublin, 1762).

contains in plenty. It could be that ephemeral publications, such as topical squibs and satires, physically flimsy and quickly out of date, were not worth noting separately even assuming that they had survived hard usage. However, Hingston's catalogue seems too comprehensive for silent omissions of this kind to be likely. Indeed, it is notable that he entitled the list as being of 'my books and *pamphlets*'.[90] From the numbering of each entry, it is evident that he noted each separate title even when several pieces were bound together into a single volume. It may be that topical tracts were passed quickly to neighbours rather than being retained for the library. At all events, the few tracts sparked by current controversies contrast with an abundance of novels and plays. About seventy titles can be identified confidently as fiction. There were printed editions of approximately 260 different plays, as well as collected editions of classics like Shakespeare.[91] Either Hingston himself, or others within his household, hungered for such printed entertainments.

How this came about can only be guessed. In the absence of revealing biographical details about the family, it is impossible to know if the published texts were bought after seeing the plays in Dublin, in the provinces for performance within the family and among neighbours, or simply because of a delight in reading them. Others from Ireland showed a similar penchant for printed plays. Earlier in the eighteenth century, the second duke of Ormond and the youthful Lord Louth, stocked their libraries with the texts of plays. The Ormonds' collection is notable for the quantity of works in foreign languages, suggesting that the plays may have been a souvenir of continental sojourns.[92] Moreover, the Ormondes divided their time between England, Ireland and the continent, so they had ample opportunities both to see, and to buy the copies of, plays. In 1723, Edward Hamilton, a Dublin bookseller, sold Louth copies of Nicholas Rowe's *Tamerlaine* (for eight pence), Thomas Otway's *Venice preserved* and *Hamlet* (for one shilling each). Louth's taste can be traced back to his school days at Greenwich

90 My emphasis.

91 The nine-volume set of Shakespeare had been published in London in 1751.

92 The National Archives: Public Record Office, London, FEC 1/877; T. Barnard, 'Introduction' in T. Barnard and J. Fenlon (eds), *The dukes of Ormonde, 1610–1745* (Woodbridge, 2000), pp. 36–7.

when he went frequently to the theatre.[93] A contemporary of Hingston, George Cockburn, a merchant and military agent, who lived in Dublin, showed similar proclivities. Cockburn's books included a set of eighteen bound volumes of plays, a further separate volume, Otway's plays, Chetwin's plays, and the eight volumes of Warburton's and Pope's edition of Shakespeare.[94]

The printed plays in Hingston's collection included many by authors, sometimes Irish in origin, popular in London: Congreve, Dryden, Farquhar, Vanbrugh and Wycherley. The works of several dramatists, notably John Dryden (and Shakespeare), were in London printings. His copies of Dryden's *Albion and Albanius*, *Almanzor and Almahide*, and *Amboyna* were published in London in 1735.[95] Possibly they had been bought during a trip to London. 1735 fell conveniently between Hingston's undergraduate studies at Dublin University and his ordination. But equally there would have been no difficulty about finding the texts in Dublin or even in Cork. Racine's *Andromache* was present in the library in an English translation published in London in 1675. In contrast there were two stray volumes (in French) from different editions of Molière's works.[96] The plays of the Munster grandee, Orrery, produced first in the 1660s, were known to Hingston through a two-volume edition issued in London in 1739. However, the majority of the editions of plays were Dublin printed. Some were published long after their original performances in London; others followed closely on the premières in either England or Dublin. Hingston had a copy of John Vanbrugh's *Aesop, a comedy*, printed in Dublin in 1725. The piece had been acted in the city during the previous year.[97] He also owned the text of Aaron Hill's tragedy *Alzira* in a Dublin printing of 1736: the year in which it was staged in London. *Alzira* is known to have been per-

93 Guardian's accounts for Lord Louth, 1709–18, s.d. 8 Feb. 1708[9], 22 Aug. 1711, NLI, MS 40,083/23; bill of E. Hamilton, 23 Apr. 1723, ibid., MS 40,081/2.

94 G. Cockburn, inventory of books, 1 Jan. 1764, British Library, Add. MS 48,314, ff. 32–3v.

95 John Dryden, *Albion and Albanius, an opera* (London, 1735) [EST C22169]; John Dryden, *Almanzor and Almahide, a tragedy* (London, 1735) [ESTC T63293]; John Dryden, *Amboyna, a tragedy* (London, 1735) [ESTC N71637].

96 The first is dated at Paris, 1679; the second, Paris, 1697.

formed in Cork in 1763, and this may have been the stimulus to find a copy.[98] Indeed, most of the plays of which Hingston detailed printed texts were performed regularly in the Dublin and Cork theatres.

By the mid-eighteenth century, opportunist Dublin publishers were quick to issue versions of the latest hits by Isaac Bickerstaffe, George Colman, David Garrick and Arthur Murphy. The texts of plays were published in the same year as they were first acted in London and in advance of their Dublin productions. Bickerstaffe's *Maid of the mill*, a comic opera, had its début at London's Covent Garden in 1765. Hingston's copy of the libretto had been published in Dublin, also in 1765. The piece was given in Cork in the same year and again in 1769.[99] Similar synchronization marked the opening of Bickerstaffe's *Lionel and Clarissa* at Covent Garden in 1768 and the publication of the text in Dublin, a copy of which Hingston had acquired. This comedy would be performed in Cork in every season between 1771 and 1774. In the case of Bickerstaffe's *Hypocrite* (an adaptation of Colley Cibber's *The non-juror*), the Dublin edition of the text, which Hingston owned, had appeared a year after the London opening in 1768. Hingston failed to mention the date of his Dublin edition of another Bickerstaffe success, *Love in a village*. It had been performed first at Covent Garden in 1762, and was given in Cork in 1765, 1766 and 1769.[100]

A second successful dramatist of the 1760s whose work appealed to Hingston was the elder George Colman. Evidently others in Ireland shared his enthusiasm. Hingston's Dublin edition of Colman's *The jealous wife* was dated 1761, the year in which it had opened at Drury Lane in London. His copies of Colman's farces, *Polly Honeycomb* and *Deuce is in him*, had both been printed in Dublin a year after their first London performances, respectively in 1760 and 1763.[101] Arthur Murphy was another favourite with the Hingstons, perhaps because of his Irish ancestry. The Irish link probably encouraged Dublin printers quickly to issue texts of his latest plays. Hingston noted Dublin-printed copies of

97 La Tourette Stockwell, *Dublin theatres and theatre customs, 1637–1820* (Kingsport, TN, 1938), p. 60.

98 W.S. Clark, *The Irish stage in the county towns, 1720 to 1800* (Oxford, 1965), p. 314.

99 Ibid., p. 296.

100 Ibid., pp. 295–6.

101 Ibid., pp. 301–2.

The apprentice (dated 1756, the year of its Drury Lane opening); *The citizen* (1761, when it was first performed at Drury Lane); *The old maid*, printed the year following its 1761 debut; *All in the wrong*, similarly appearing a year after its Drury Lane première in 1761; and *What we must all come to*, its Covent Garden first performance and Dublin edition both in 1764.[102]

Not all approved of theatrical amusements. Qualms about the appropriateness of this taste in a cleric, or clerical household, are aired in tracts owned by Hingston. Whatever misgivings were felt, they did not scotch the Hingstons' passion for the library was rich in drama. It seems likely that the books were enjoyed by the family rather than just its head. Some plays may well have been read aloud and provided the basis of entertainments in which neighbours joined.

Novels, although not as numerous as the play books, were nevertheless well represented. Again the dependence of Anglophone readers in Ireland on what had been written and published originally in England was clear. Hingston recorded a four-volume edition of *Pamela*, a three-volume *Tom Jones*, three volumes of *Tristram Shandy*, two volumes of *Humphrey Clinker*, two volumes of *The sentimental journey*, and another two of *Roderick Random*. All had Dublin imprints.[103] There was even a compendium of twelve novels, published in Dublin in 1760. More puzzling is *Mrs Tenducci, her letters*, printed in Cork in 1757.[104] The elopement of Dorothea Maunsell, the daughter of a Limerick notable, with Giusto Tenducci, a castrato performing with theatrical companies in Ireland, proved a sensation. It led quickly to an embellished account of the episode. However, it occurred in 1766. The earlier Cork account is not otherwise known. It may possibly be a spurious imprint, intending to exploit the prurience of readers in south Munster. Even so, a date of publication nine years before the celebrated event makes little sense. It could be that Hingston made a mistake in detailing the volume. But, since his standard of bibliographical accuracy is

102 Ibid., pp. 323–4; Stockwell, *Dublin theatres and theatre customs*, pp. 216–17.

103 R.C. Cole, *Irish booksellers and English writers, 1740–1800* (London, 1986), pp. 62–86; Pollard, *Dublin's trade in books*, pp. 88–9.

104 No. 700 in Hingston's catalogue.

generally high, it is unlikely that he was incorrect in recording both the date and place of publication.[105]

Occasional items bridged what has come to be regarded as the gulf between polite and demotic literature. These included *The seven champions of Christendom* in a Dublin edition of 1716, and the more recent Dublin printings of *Valentine and Orson* (1771) and of *The Seven wise masters* and *Mistresses*.[106] Also present was the account of the highwayman, James Freney, whose exploits had recently titillated the law-abiding, especially in Munster, the scene of his depredations.[107] The genre was further represented in the Dublin printed *History of Irish highwaymen*.[108] It is impossible to know whether the sensationalist tales, and the considerable weight of plays and novels, represent Hingston's own liking for lighter reading or the fancies of the younger and female members of the household.

V

The ample supply of novels and plays is linked with another feature of the library that will merit more investigation: the proportions of works printed in Dublin or the Irish provinces, in England and further afield. The total of Irish provincial printings was overshadowed by the productions of Dublin, London and the European continent. In the main, the volumes printed on the continent were older books, published before 1700, typically classics, theology and history. Probably they had been in Ireland for some time before they were picked up by the Hingstons. The upheavals of the seventeenth century enabled the interested to buy cheaply many curiosities and rarities formerly in the libraries of

105 *Mrs Tenducci, her letters,* duodecimo (Cork, 1757); *A true and genuine narrative of Mr and Mrs Tenducci* (London, 1768) [ESTC 117835]; T.J. Walsh, *Opera in Dublin, 1705–1797: the social scene* (Dublin, 1973), pp. 134–9. One example of an error is the second reference to MacGeoghegan's *History,* which is indexed under both title and author. The second entry gives the imprint correctly as Paris, but the date as 1738, rather than 1758, which is correct and so given in the first reference.

106 R. Gillespie, *Reading Ireland: print, reading and social change in early modern Ireland* (Manchester, 2005), pp. 90, 92, 162–4; Ó Ciosáin, *Print and popular culture in Ireland*, pp. 72–83; Pollard, *Dublin's trade in books*, pp. 34, 63–4, 192, 221–3.

107 *The life of James Freney* (Dublin, 1770).

108 Ó Ciosáin, *Print and popular culture in Ireland,* pp. 84–99.

Catholics (or indeed of Protestants when ransacked by insurgents). Auctions, specialist dealers, gifts and even plunder were sources of supply. In contrast, the recent dates of many of the works published in Ireland – notably the novels and plays – make it more likely that they were bought new by the Hingstons. Without any accounts, bills or correspondence, it cannot be shown whether the new books were bought when family members visited Dublin, Cork or Limerick, were ordered on the strength of advertisements in the newspapers and on personal recommendations, or supplied by country shopkeepers and peddlers.

Despite the presence of books published in continental Europe, the library provides no clear evidence that the Hingstons had travelled beyond Ireland and Britain. It did contain a few works relating to Italy: for example, a folio of views of Venice, which had been printed in London. The Venetian theme is reprised in one among his small group of manuscripts: a copy of Amelot de l'Houssaye's 'Government of Venice'.[109] Hingston recorded two editions of Guicciardini's *History of Florence* and the 1595 London printing of Macchiavelli's *The Florentine history*.[110] Then, too, there was Giannone's *Civil history of Naples.* The ownership of these volumes hardly betokens a journey to Italy. But the presence of Richard Lassels' *Voyage d'Italie* and the recent work by Samuel Sharp, *Letters from Italy*, may be connected with a desire to go there, if not a souvenir of trips.[111] One other work suggestive of a foreign foray is the printed description of the rarities in the anatomy hall in Leiden: a popular stop for educational and recreational tourists from Britain and Ireland in the early eighteenth century.[112] Armchair travel was satisfied by histories of Jamaica and Siam.[113]

109 Since an English translation had been published in London in 1708, it is curious that Hingston had the manuscript.

110 F. Guicciardini, *History of the wars of Italy* (London, 1599); idem, *History of the wars in Italy* (Dublin, 1754); (Hingston recorded it as London, 1754, in six volumes. There is a ten-volume edition London, 1753–6 [ESTC S17853]); N. Macchiavelli, *The Florentine history* (London, 1595).

111 Paris, 1671; Dublin, ?1767, but undated in Hingston's list: [ESTC T134845].

112 *A catalogue of all the chiefest rarities in the Publick Theater and Anatomie-Hall, of the University of Leyden* (Leiden, 1701); Barnard, *Making the Grand Figure*, pp. 316, 320, 321.

113 [Charles Leslie], *A new history of Jamaica* (Dublin, 1741); Simon de la Loubère, *A*

According to his own calculations, forty-two of Hingston's titles had been published in Cork, eight in Limerick, six in Belfast, two in Newry and one each in Drogheda and Clonmel. The Cork works included several of local interest or written by inhabitants of the city, such as the dean of the cathedral, Rowland Davies, William Boles (alias Alexander the Coppersmith) and Dr Patrick Blair.[114] George Russel, whose *Works* were issued in 1769, was Church of Ireland incumbent in the West Cork parish of Schull.[115] The 1721 edition of Richard Mead's *Discourse on plague* was occasioned by an outbreak of the disease in the city. *The trial of [HRH] the duke of Cumberland for criminal conversation with Lady Harriet G-r* cashed in on a sensation of the season.[116] A few in Hingston's assemblage of plays and novels had been printed in Cork: Madame d'Aulnoy's *The history of Hypolitus, the earl of Douglas*;[117] Henry Dell's *Minorca, a tragedy*; Samuel Foote's *The West Indian*; Richard Brinsley Sheridan's *The governess* and Musgrave Heighington's *The enchanter*.[118]

Limerick publications showed the same mixture. *A history of the port* by John Ferrar, an account of the municipal dispute stirred up by Charles Massy, the dean of Limerick, a military history by an officer of the garrison and an anti-Catholic diatribe were balanced by Richard Bulkeley's *Poems* of 1739 and the most ambitious production, forty volumes of the *Magazine of Magazines*, produced in the city in 1771. The

new historical relation of the kingdom of Siam (London, 1693).

114 For Cork imprints see Dix, 'List of books, etc. printed in Cork', *JCHAS*, 6 (1900), pp. 168–174, 233–40; 7 (1901), pp. 104–110; 9 (1903), pp. 97–105, 264–8; 11 (1905), pp. 24–6; H. Fenning, 'Cork imprints of Catholic historical interest, 1723–1804', ibid., 100 (1995), pp. 129–48; also, J.P. McCarthy, 'In search of Cork's collecting traditions', ibid., 100 (1995), pp. 29–46.

115 George Russel, *Works*, 2 vols (Cork, 1769); Dix, 'List of books, etc. printed in Cork', *JCHAS* 7, p. 108; 11, p. 266; M. Kennedy, 'At the Exchange: the eighteenth-century book trade in Cork' in C. Benson and S. Fitzpatrick (eds), '*That Woman!': studies in Irish bibliography: a festschrift for Mary 'Paul' Pollard* (Dublin, 2005), p. 153. The long list of subscribers does not include Hingston.

116 Cork, 1770.

117 This is advertised by William Flyn in Russel, *Works*, at the end of volume two.

118 Dix, 'List of books, etc. printed in Cork', *JCHAS* 6, p. 239; 11, pp. 101, 102, 105; B. Boydell, *A Dublin musical calendar* (Dublin, 1988), p. 299; Kennedy, 'At the Exchange', pp. 151–2.

last was a version of the popular London serial, *The Universal Register*, supplemented with some Irish material. Its publication in Limerick suggests both a growing market in south Munster among those of Hingston's means for international, British and local news, and the opportunism of printers in meeting the demand.[119]

Belfast chiefly supplied the Hingstons with entertainments. Its products included a 1741 printing of the *Scots Hudibras,* 'a dramatic satire', *The toyshop* by Robert Dodsley, and three farces: *The cobbler of Preston* (1751); *Sir John Cockle at Court* (1748); and *The miller of Mansfield* (1750).[120] The two Newry printings were similar: Forde Brownlow's *The miraculous cure, a farce* (1771) and *The ballad of Robin Hood* (1769).[121] The pattern is repeated with a 1754 Glasgow edition of a comedy, *The provoked husband,*[122] and the solitary and hitherto unknown Drogheda production, *Coffee house jests of 1760.*[123] Finally, the Clonmel publication was *The swaddler*, Amyas Griffith's farce issued in 1771.[124] There was one Boston imprint: *Essay on gratitude* (1732). What prompted provincial printers to produce particular works can, for the moment, only be guessed. Hope of profit is the likeliest explanation. It

119 It is not mentioned in either R. Munter, *The history of the Irish newspaper, 1685–1760* (Cambridge, 1967) or M. Kennedy, 'Eighteenth-century newspaper publishing in Munster and south Leinster', *JCHAS*, 103 (1998), 67–88. One surviving copy is recorded – in Yale University Library – which consists of only twenty-four volumes. ESTC P2507.

120 Charles Johnson, *The cobler of Preston* (Belfast, 1751) [ESTC 167086]; Robert Dodsley, *Sir John Cockle at court* (Belfast, 1748) [ESTC T217621]; Robert Dodsley, *The toy-shop* (Belfast, 1747) [ESTC T175172]. Neither Samuel Colvill's *Scots Hudibras* nor Dodsley's *The miller of Mansfield* is known in these Belfast editions. But see: J.C. Greene, *Theatre in Belfast, 1736–1800* (Bethlehem and London, 2000), p. 48. The latter farce had been acted in Dublin in the 1748–9 season. E.K. Sheldon, *Thomas Sheridan of Smock-Alley* (Princeton, 1967), p. 440.

121 Brownlow's farce is ESTC N6093. *Robin Hood's garland* is known in Newry edition of 1794: [ESTC N65570].

122 There were other Glasgow printings in the collection: for example, editions of Cicero, dated 1691 and 1732 and *Civilis imperii origo et causae* (1750).

123 J. Finegan, 'Georgian Drogheda and the printed word', in G. Long (ed.), *Books beyond the Pale aspects of the provincial book trade in Ireland before 1850* (Dublin, 1996), pp. 33–50.

124 ESTC 161097. Three surviving copies of this publication have been located. See, too: A. McClelland, 'Amyas Griffith', *Irish Booklore*, 2 (1972), 6–21.

may be possible to show that the plays had recently been performed by touring companies in the towns in which versions were then printed. It could also be that the opportunists simply inserted their own title pages before sheets that had been printed in Dublin or England.

Hingston's catalogue adds tantalizingly to what is known about the output of Irish printers. Exhaustive analysis will show how many of his volumes were in hitherto unknown editions. Other questions – the circumstances in and the reasons for which the books were acquired and about their uses – are unlikely to be answered so confidently. The clerical vocation, the education of family, neighbours and parishioners and utilitarian needs explain much in the collection. Yet, pleasure and diversion are palpable. The froth of novels and plays, together with the *Magazine of Magazines*, lightens the stodgy theology, history and philosophy. Notable absences are lavish (and therefore costly) volumes, illustrated with plates depicting ancient and more modern structures. Few of these were produced in Ireland before 1774. Often Irish owners acquired them when travelling outside the island, as mementoes of foreign tours. Again, it seems that the Hingstons were not in a condition to send offspring on such travels, although among the swells of Irish Protestant society continental tourism was almost de rigueur.

In the end, the obvious impression from the catalogue is of bulk as well as of diversity. A library of approximately 2,500 titles in a country parsonage is startling. The questions of how many of the tomes or pamphlets were read, and, if read, how understood, remain insoluble. At the very least, the miscellaneous nature of the Hingstons' books suggests powerfully that studiousness was balanced by vivacity, and the potentially dour and drab by the light-hearted and frivolous. Many of the novels, plays and poems carried moral cautions, but so sweetly sugared as also to divert and delight.

Publishing the law

John Finlay, 1780–1856

W.N. OSBOROUGH

IN 1830, JOHN CUMMING, the bookseller and publisher with an address at 16, Lower Ormond Quay in Dublin, advertised for sale amongst his stock of legal titles no less than six books written by the same author, one John Finlay. The order in which these six volumes had been published is not entirely clear, the existence of several editions for individual titles helping to compound the problem over establishing that order. It is thought, however, that the sequence of law books from this prolific early nineteenth-century Irish author started with his *Office and duty of church-warden and parish officer in Ireland*, which appeared in 1824, if not before. A second edition is recorded for 1827 and a third for 1830. *Office and duty of church-warden* was followed in 1825 by *The law of landlord and tenant for Ireland*, which went into a second edition in 1835, and in 1827 by *A treatise on the laws of game, and inland fisheries in Ireland*, and, in 1828, by *A treatise on the law of tithes in Ireland, and ecclesiastical law, connected therewith*. The next year, 1829, there came *A treatise on the law of renewals, in respect to leases for lives renewable for ever, in Ireland, with the cases and decisions thereon, including an extensive report of the important and heretofore unpublished case of Fitzsimon* v. *Burton; also the cases in respect to the renewals of ecclesiastical leases in Ireland, with tables for the calculation of fines*. There followed, in 1830 itself, the first ever digest of law reports in Ireland, *A digest and index of all the Irish reported cases in law and equity, from the earliest period to the present day, and also of the reported cases in ecclesiastical and criminal law, together with a variety of original cases from authentic sources, containing many of the judgments of Lord Lifford, Chancellor.*

John Cumming was a well-established figure in the early nineteenth-century Dublin book trade. He entered into a succession of partnerships and profited from his position as the Irish agent for a number of English publishers. Three years before his death, however, he was to be declared bankrupt. The books he sold at his Lower Ormond Quay premises were a veritable miscellany, as is clear from a newspaper advertisement of 1815 listing *A catalogue of French, Italian and other foreign books, recently imported from Paris.*[1] Cumming's flyer of August 1830 demonstrated his interest in the developing market for law books. It is instructive that it acknowledged the productions of two other Irish legal authors of the times, apart from John Finlay. The flyer thus announces a single volume from George Robinson: *Election law: the laws regulating the registry of freeholders, and proceedings at contested elections in counties*; and no less than four volumes from W.C. McDermott – *The criminal code for Ireland*; *A treatise on tolls and customs of Ireland*; *A compendious digest of the statutes passed in the session of 1828, for Ireland*; and *A digest of the statutes passed in the 3rd session of Parliament of 1829, for Ireland.* It is, however, the six very varied volumes produced by the first of our authors which stand out in the history of Irish legal publishing for the period. By any measure, and certainly as regards both the range of the subject matter and the quality of the treatment accorded the themes that are the focus of attention in the first five titles, this was an outstanding achievement.

The author of all these works was the Irish barrister John Finlay. No satisfactory likeness of Finlay survives, but, in 1833, he posed for Augustin Edouart (1789–1860), who did a silhouette – one of six thousand or more 'portraits' Edouart did during his extended Irish sojourn of 1833 to 1839.

1 For these details on Cumming, I am grateful to Charles Benson, Keeper of Early Printed Books at Trinity College, Dublin Library.

1 Silhouette of John Finlay by Augustin Edouart,
courtesy of the National Library of Ireland

This survives in the Joly collection in the National Library, and has a signature in Finlay's own handwriting (Fig. 1).[2]

Finlay was born in Londonderry in 1780, the first son of Jeffrey Finlay; his mother was an Anastasia Furlong. Educated at Trinity College Dublin, Finlay later decided on a legal career. Admitted as a student of King's Inns in 1805 and, under the then obligatory rules, of the Middle Temple in 1806, he was finally called to the Irish bar in January 1809. We know from his King's Inns admission papers that at the time of his entry there as a student, his father was already dead.[3] Of the rest of Finlay's family little is known. One detail is extant, however: his sister Eliza, who survived him by thirteen years, is buried in the same grave as her brother in Glasnevin cemetery in Dublin.[4] Finlay was later to marry.

2 W.G. Strickland, *A dictionary of Irish artists*, 2 vols (Dublin and London, 1913), i, pp. 314–17. R.M. Elmes, *Catalogue of engraved Irish portraits mainly in the Joly collection* (Dublin, 1937), p. 72.

3 *King's Inns admission papers, 1607–1867*, ed. Edward Keane, P.B. Phair and T.U. Sadleir (Dublin, 1982), p. 163.

4 W.J. Fitzpatrick, *History of the Dublin Catholic cemeteries* (Dublin, 1900), p. 72.

He was also to acquire an LL D degree about the time of his first judicial appointment.

Finlay's progress through the ranks of the Irish judiciary can be briefly rehearsed. A view, held by one nineteenth-century commentator, was that the recognition which promotion to the bench plainly evinced had, in Finlay's case at least, been unduly delayed.[5] However this may be, in 1826 or 1827, Finlay, now in his forties, was appointed seneschal of Kilmainham, one of the manorial jurisdictions then extant in Dublin. He was to retain this position until 1854, just five years before all Irish manor courts were abolished.[6] As seneschal of Kilmainham, Finlay is unlikely to have shouldered anything approaching a heavy judicial burden. In 1831, additional preferment came his way which, however, promised rather more to do. This was when he was made assistant barrister for County Leitrim – a promotion that necessitated frequent travel to the county for the discharge of legal business there, notably to dispose of all civil bills lodged in connection with ordinary civil litigation. The post of assistant barrister had been established in 1787, and the years since the Union had witnessed a substantial increase in the assistant barrister's jurisdiction, covering such matters as ejectment and replevin. Not long after Finlay's appointment, by virtue of a measure linked to changes in the franchise introduced by the great Reform Bill of 1832, assistant barristers were empowered to deal with the often controversial question of the registrability of voters.[7] Finlay stayed as assistant barrister for Leitrim until 1839 when he was appointed assistant barrister for Kerry – a transfer which, of course, involved very much longer journeys down to the south-west. In 1843, Finlay was moved again – to Roscommon. Finally, in 1849, he was made assistant barrister for County Louth, a position he retained until he retired in 1854, a few short years after assistant barristers had had added to their responsibilities yet another fresh jurisdiction, one in insolvency.[8] From 1835 to 1837 Finlay held another

5 Ibid., p. 71.

6 By the statute 22 Vict., c. 14. See further W.N. Osborough, *Studies in Irish legal history* (Dublin, 1999), pp. 255–6.

7 Osborough, *Studies*, pp. 253–4.

8 By the statute 14 & 15 Vict., c. 57. The details of Finlay's judicial career have been gleaned from successive Dublin directories.

judicial post: that of surrogate to Sir Henry Meredith, the judge of the Irish court of Admiralty. It is worth noting that Finlay was never elevated to the ranks of king's or queen's counsel, nor was he ever elected a bencher of King's Inns.

No evidence of Finlay's work as seneschal of Kilmainham, as Meredith's surrogate in Admiralty or as the assistant barrister for the various counties to which he was successively appointed is immediately accessible. Any reputation to which as a legal figure he may be entitled is due exclusively to his published writings.

In Finlay's case, it seems clear enough that his *cacoethes scribendi* originated in his liberal politics and in his close association as a young man with the Catholic cause – not perhaps the most customary of involvements at the time for someone brought up in the Church of Ireland. Finlay, it was to be claimed, 'had been a thinker almost from the cradle, and a philosopher in the nursery'.[9] An extract from his address as auditor of the Trinity College Dublin Historical Society in 1809, and reproduced by W.J. Fitzpatrick, shows something of Finlay's oratorical abilities as well as of his political sentiments. 'Tremble the wretch', he apparently was to assert,

> who, in the mask of hypocrisy, hopes to deceive her searching eye, and escape the grave, eluding punishment or evading detection; let him tremble at the certainty that history at last plucks off the vizor of the villain, and that not even the tomb can afford concealment to guilt, or shelter to the guilty, when History is the prosecutor and Posterity the judge.[10]

Why exactly Finlay was attracted to the Catholic cause is impossible to establish. It is enough to know that he was – a political stance well enough attested to by the first publication of Finlay of which there is a

9 Fitzpatrick, *Dublin Catholic cemeteries*, p. 71.

10 Quoted, ibid. For the complete address see John Finlay, *Miscellanies: the foreign relations of the British Empire; etc.* (Dublin and London, 1835), ch. 3. The address had earlier been published as a separate pamphlet by T. Burnside of 87 Dame Street: *Speech of John Finlay, esq., delivered from the chair of the Historical Society at the commencement of the present session* (Dublin, 1810).

record. In 1812, there was published in Dublin *A report of Mr Finlay's admirable discussion on the question of the veto delivered by him at Galway, 26 July 1812*. In December the following year, Finlay, by now a leading light on the Catholic Board, proposed the motion to set aside a substantial sum of money for a gift to Daniel O'Connell in recognition of his stalwart endeavours for the Catholics of Ireland. Finlay's eulogy, spread over seven columns in the *Dublin Evening Post*, concluded with the proposal:

> That a service of plate, of the value of One Thousand Guineas, be presented to Dan O'Connell, esq., on the part of the Catholic people of Ireland as a small tribute of their gratitude for the unshaken intrepidity, matchless ability and unwearied perseverance with which, in despite of Power and Intolerance, he has uniformly asserted the Rights, and vindicated the calumniated Character of his Catholic Countrymen.

Many years later, in 1835, when this service of plate eventually came to be inscribed, O'Connell, having been asked to supply the date when the Catholic Board's motion had been moved, admitted he could not remember the precise date, but went on to advise a correspondent to consult the newspapers, or better still, to ask Finlay.[11]

Finlay's intimacy with O'Connell endured for many years. In 1819 and 1820, for example, Finlay and his wife would dine with him, or he with them.[12] Affairs in South America in 1823 generated letters to O'Connell who would be desired specifically to hand on greetings to Finlay.[13] In 1824, both men would meet at charity functions and public dinners to aid the cause of Catholics long pressing for Emancipation.[14] A mutual friend throughout these years was another liberal Irish

11 *The correspondence of Daniel O'Connell*, ed. Maurice O'Connell, 8 vols (Dublin, 1973–80), v, 344; *Correspondence of Daniel O'Connell the Liberator*, ed. W.J. Fitzpatrick, 2 vols (London, 1888), ii, pp. 43–4; *Dublin Evening Post*, 16 Dec. 1813.

12 *O'Connell corr.*, ed. M. O'Connell, ii, 220, 238.

13 Ibid., ii, pp. 255, 339, 406, 468. John Devereux was writing from places such as Bogota and Caracas.

14 *O'Connell corr.*, ed. M. O'Connell, iii, pp. 43, 220.

Protestant, Charles Phillips, like Finlay, a graduate of Trinity College, Dublin. Phillips, who was born around 1787 in Sligo, graduated from Trinity in 1806. He was called to the Irish bar in 1812 but made his career eventually in England. Called to the English bar in 1821 (like Finlay his allegiance had been to the Middle Temple), Phillips became a bankruptcy commissioner in Liverpool in 1842, before transferring to London four years later. Phillips died in 1859, three years after Finlay, his erstwhile friend and colleague.[15]

In 1817, Finlay penned the preface to *The speeches of Charles Phillips, esq., delivered at the bar, and on various public occasions, in Ireland and England*. Here Finlay mounts a vigorous defence of oratory, or rather of oratory as declaimed by his friend Phillips:

> There are few men without some judgment, but there are many men without any imagination: the latter class never did, and never can produce an orator. Without imagination, the speaker sinks to the mere dry arguer, the matter-of-fact man, the calculator, or syllogist, or sophist: the dealer in figures; the compiler of facts; the mason, but not the architect of the pile: for the dictate of the imagination is the inspiration of oratory, which imparts to matter animation and soul.[16]

It is probable that it was a passage such as this that was to lead Fitzpatrick to state that he preferred to recall Finlay as a 'miscellaneous writer' rather than as an author of books on law[17] – and, indeed, this remains a defensible position.[18]

15 *Oxford dictionary of national biography*, 60 vols (Oxford, 2004), xliv, pp. 99–100.
16 *Speeches of Charles Phillips ...* (London, 1817), preface, p. x.
17 Fitzpatrick, *Dublin Catholic cemeteries*, p. 71.
18 John Finlay's *Miscellanies* (above, n. 10) cover politics, foreign affairs, biography and theatre criticism. Other publications whether signed by Finlay or attributed to him testify to an undoubted versatility. These include: (i) Anon., *A statement of the objects of the Association for the Improvement of Prisons, and of Prison Discipline, in Ireland; together with some reports of particular gaols in that Kingdom* (Dublin, 1819) (publisher: Christopher Bentham, 50 Stephen Street); (ii) John Finlay, *Letters addressed to the Irish government, on local taxes, the Irish collieries, etc. etc.* (Dublin, 1822) (publisher: Christopher Bentham, Eustace Street); (iii) Anon., *Statement of*

In *The speeches of Charles Phillips* it is the content of certain of Phillips' own speeches that enables us to put together an image of Finlay in his days as the young progressive barrister. In an address to the Catholics of Sligo, Phillips specifically names Finlay as a fellow victim of general Protestant prejudice.[19] 'Every calumny', Phillips was to remark,

> which the venal, and the vulgar, and the vile, are lavishing upon you is visited with exaggeration upon us. We are called traitors, because we would rally round the crown an unanimous people. We are called apostates, because we will not persecute Christianity. We are branded as separatists, because of our endeavours to annihilate the fetters that instead of binding, clog the connection.[20]

Subsequently, in a Dublin speech, Phillips uttered sentiments with which, again, Finlay undoubtedly would have agreed. The code against which the Catholics had been petitioning, Phillips began, is 'a vile compound of impiety and impolicy'. Many Irish are British diplomats, he remarked to his audience:

some of the causes of the disturbances in Ireland, and of the miserable state of the peasantry; with a plan for commencing on sound principles an amelioration of their condition, thereby removing the causes of the disturbances, and bringing the country into a state of peace and quietness (Dublin, 1825) (publisher: Bentham and Hardy, Cecilia-Street); (iv) 'Eight seven', *Prosperity of the labourer the foundation of universal prosperity* (Dublin, 1827) (publisher: Joshua Porter, 72 Grafton Street). The attribution of the authorship of (i), (iii) and (iv) to Finlay is made in a catalogue in the King's Inns library. I am grateful to the Librarian there, Mr Jonathan Armstrong, for his assistance.

19 Something else that had brought Finlay and Phillips together was their falling out with the Dublin publisher, John Magee, junior: *O'Connell corr.*, ed. M. O'Connell, ii, 102. The general background to press prosecutions of the period – the context in which this falling out took place – is explained in Brian Inglis, *The freedom of the press in Ireland, 1784–1841* (London, 1954), pp. 136–41. And for a detailed account of the prosecution of Magee for seditious libel over an intemperate article appearing in the *Dublin Evening Post* and which Finlay is claimed to have written, see Denis Gwynn, *Daniel O'Connell* (Cork and Oxford, 1947), pp. 98–107.

20 *Speeches of Charles Phillips*, p. 3.

> At this instant, is there a country in the world to which England has not deputed an Irishman as her representative? She has sent Lord Moira to India, Sir Gore Ouseley to Ispahan, Lord Stuart to Vienna, Lord Castlereagh to Congress, Sir Henry Wellesley to Madrid, Mr Canning to Lisbon, Lord Strangford to the Brazils, Lord Clancarty to Holland, Lord Wellington to Paris.[21]

Later in the same speech, Phillips confronted head-on the consequences for individual Catholics of the denial prior to 1829 of Emancipation. Again, Finlay can be surmised to have been in entire agreement:

> Though genius flashed from his eye, and eloquence rolled from his lips; though he spoke with the tongue of Tully, and argued with the learning of Coke, and thought with the purity of Fletcher, he can never rise – he is a Catholic! Merciful God! What a state of society is this in which thy worship is interposed as a disqualification upon thy Providence.[22]

Finlay qualified as a barrister in 1809. Substantially later he obtained an LL D, giving him the entitlement also to practise as an advocate in the ecclesiastical courts and in Admiralty. Largely self-taught in ecclesiastical law, he had earlier been asked to offer professional advice to parishioners in three Dublin city parishes: St Mark's, St John's and St Nicholas' Without. Groups within each of these parishes questioned the validity of applotments raised in 1816 and 1817[23] in connection with minister's money – the tax levied in certain corporate towns in support of the established clergy by way of substitution for tithe. The texts of these opinions which Finlay furnished are included in the appendix to Finlay's

21 Ibid., p. 60.

22 Ibid., p. 64. Tully is Cicero. Fletcher is the judge of that name appointed to the Irish court of Common Pleas in 1806. Fletcher had practised as a civilian whose merits were also to be extolled by Finlay in his *Office and duty of church-warden and parish officer in Ireland*. Fletcher registered a partial dissent in the case of *Taaffe* v. *Downes* (1812–13).

23 No date is actually supplied by Finlay for the request for an opinion emanating from the parish of St Nicholas' Without.

book on the office of churchwarden.[24] There the problem at St Mark's is dealt with at the greatest length.[25]

The complaining parishioners had forwarded to Finlay the complete draft applotment dated October 1816.[26]

	£	s.	d.
Senior curate early service	100	0	0
Parish clerk	5	13	9
Parish clerk in lieu of his Nine pences	27	10	0
Vestry clerk	25	0	0
Ditto, ditto: quills and paper	1	14	1 1/2
Organist's salary and instructing children	40	0	0
Sexton	40	0	0
Collector	50	0	0
Engine keeper	11	7	6
Beadle's salary and clothes	28	8	9
Gallery sextoness	13	13	0
Vestry maid	4	11	0
Bell-ringer	8	0	0
Grave digger	10	0	0
Medicines for poor	34	2	6
Coffins ditto ditto	40	0	0
Coals for church	5	0	0
Bread and wine for sacrament	15	0	0
Tuning the organ	5	13	9
Printing advertisements, etc., etc.	20	0	0
Slates and repairs of roof	10	0	0
Glazing windows	5	0	0
Ironmonger's account and repair of stoves	10	0	0
Engine repairs	5	0	0
Incidental expenses	25	0	0
TOTAL	540	14	4 1/2

24 *Office and duty of church-warden*, pp. 190–204.
25 Only the quantum of the draft applotment – £348 – is given. St John's complained it was the poorest and smallest of the parishes.
26 *Office and duty of church-warden*, p. 190.

Some years before, Judge Radcliffe, judge of the court of Prerogative and Faculties – the senior Irish ecclesiastical court – had furnished an opinion which, if it was followed, cast doubt on the validity of such an applotment as was raised at St Mark's in 1816.[27] Radcliffe took the view that only in the case of the parish of St Andrew's could certain of the items featured in the applotment above be regarded as valid. That was because an Irish statute of the Restoration period specifically authorized the churchwardens in St Andrew's to raise funds 'for the relief of the poor, and other pious uses'.[28] No such specific entitlement was bestowed upon the parish of St Mark's by legislation in the reign of Anne.[29] Radcliffe concluded, accordingly, that the levying on parishioners in St Mark's of moneys 'for the relief of the poor, and other pious uses' was illegal.

Despite the authority that necessarily attached to any opinion of Radcliffe's, Finlay in 1816 was to take a different view. The parish of St Mark's had been partly formed out of the parish of St Andrew's under the later legislation,[30] and from that he reckoned it could be inferred that the power given the churchwardens of St Andrew's had also been inherited by the churchwardens of St Mark's.

Other practical difficulties linked to the administration of Dublin city parishes came in due course to the attention of Finlay. These included the question of the right of Catholics residing in a particular parish to vote at the election of parish officers.[31] In 1804, following the death of John Gaskin who had served as the sexton at St Andrew's, a row over this very question broke out. Opinions from William Fletcher, soon to

27 Finlay only furnishes the gist of this opinion. In the preface tribute is paid to Radcliffe as a civilian: *Office and duty of church-warden*, preface, p. v.

28 The statute 17 & 18 Chas. II, c. 7: see s. 4. That was 'An act for provision of ministers in cities, corporate towns, and making the church of St Andrews, in the suburbs of the city of Dublin, presentative for ever'.

29 The statute 6 Anne, c. 21: 'An act for dividing the several parishes of Saint Andrew's, Saint Nicholas without the Walls, and the united parishes of Saint Katherine's, Saint James, and Saint John's of Kilmainham, and for making or erecting a new parish, by the name of the parish of Saint Anne, on the ground lying between Grafton-street and Merrion-street, all situate, lying, and being in or neere the suburbs of the city of Dublin'.

30 6 Anne, c. 21, s. 2. The marginal note reads 'Boundaries of St Mark's'.

31 See *Office and duty of church-warden*, pp. 177–83.

be elevated to the judgeship in Common Pleas, and from Arthur Browne, then the prime serjeant,[32] both argued that Catholics were eligible to vote for the sexton. The clergy at St Andrew's, possibly succumbing to pressure, had postponed the convening of the critical meeting. A question asked of the two lawyers demanded to know the remedy available to liberal Protestants within the parish if procrastination on the part of the incumbent and his curates continued. They also wanted to be apprised of their remedies at law if the chairman at any eventual meeting declined to allow voting by Catholics. Arthur Browne concluded his opinion of 27 August 1804 with a cautionary remark. Having restated the generally held belief that Catholics could not vote for churchwardens and parish clerks (but could for sextons), he went on:

> There undoubtedly is a certain awkwardness in the persons of one persuasion in religion electing the officers appertaining to the celebration of divine worship by those of another communion; but it is the duty of a lawyer giving his opinion, not to judge of the propriety of laws, but to declare what he apprehends the law to be.[33]

As regards the remedy available to the parish faction that had sought his advice where the clergy delayed the convening of any election meeting, Browne indicated that a writ of *mandamus* should be applied for. That, however, he accepted, might not lead to success, citing the extraordinary case reported in 1 Strange at p. 636 (a mistake, I think, for a case reported in these reports at p. 536.)[34] In further opinions, Radcliffe at some point was asked about claims to exemption from minister's money advanced on behalf of the Theatre Royal and of warehouses opened for business on sites where dwelling houses had once stood.

These various problems came to the knowledge of Finlay in a professional capacity. It is hard to avoid the conclusion that he perceived

32 Arthur Browne (1756?–1805) was also an M.P. and regius professor of laws at Trinity College. See further the entry in the *Oxford DNB*, viii, p. 149.

33 Reproduced by Finlay in *Office and duty of church-warden*, p. 182.

34 *Dominus Rex* v. *Decan' et capitul' Dublin*, 1 Strange 536; 93 Eng. Rep. 685.

the need to set them down in a format which would ensure a continuing awareness that they had once existed and had been dealt with. Thus the idea took shape of writing *Office and duty of church-warden and parish officer in Ireland* – what was to prove to be the first of Finlay's many volumes. Like Vernon and Scriven when they brought out the first of the modern sequence of Irish law reports in 1787,[35] Finlay saw the need to justify what he was at. This he proceeded to do in the preface to *Office and duty of church-warden*. In retrospect, we can see that it was a predictable enough form of *apologia*. 'The law of Ireland', Finlay wrote,

> is in many cases peculiar, and few books on legal subjects have been published in this country. Thence it happens, that an individual in Ireland is often obliged to resort for necessary information on a subject of an important yet local nature, to some treatise of English law, which must in many cases mislead the inquirer.[36]

The identical point is made, with pointed emphasis, in another of Finlay's prefaces, that which introduces his *Treatise on the law of tithes* (1828). This latter volume is perhaps the most impressive of Finlay's law books. One reason is his excellent grasp of the history of the topic.[37] As early as his address to Trinity College's Historical Society in 1809 Finlay had perceived the need for the lawyer to possess an acute historical sense. 'Of all sorts of human knowledge', Finlay had then asserted,

> historical knowledge is chiefly necessary to the legal advocate. My Lord Bolingbroke calls it the very 'vantage ground of law'. It leads you to an acquaintance with the *jura naturae*, the *leges*

35 Osborough, *Studies*, pp. 200–1.

36 Preface, p. iii. Finlay's decision to tackle the legal duties of churchwardens probably owed something to his acerbic comments on local, and especially Dublin, taxes in his *Letters addressed to the Irish government, on local taxes, the Irish collieries, etc. etc.* (Dublin, 1822). An itemized bill for such taxes over a 12-month period is furnished in Letter VIII.

37 For further enlightenment, see Maurice J. Bric, 'The tithe system in eighteenth-century Ireland', *Proceedings of the Royal Irish Academy*, 86 (1986), sect C, 271.

> *legum*, the abstract nature of all law, and indeed that book so appropriately termed the Spirit of Laws, is nothing else than a collection of inductions from the history of all nations.[38]

Whether indebted to Montesquieu or not, Finlay on tithe is a magisterial performance. Such mysteries as customary moduses, the episcopal fourth, the exemption of abbey lands, agistment tithe and minister's money are all elucidated. There is a Finlay specialty besides: actual law cases reported at length are conveniently included. Cumming's flyer of 1830 listed the further attractions:

> Also, a comprehensive explication of the Three Statutes on Tithe Composition, with the decision of King's Bench and Privy Council thereon. Leases of Tithes, by Statute, in Ireland. Forms of Leases under the Tithe Composition Acts; with Tables of the Averages. The whole being interspersed with a continuous explanation of the numerous distinctions between the law of England and Ireland on these subjects, and the latest decisions.

Finlay himself understood the marketability of the volume. In the preface, he wrote bluntly:

> The necessity of a work of this kind is obvious; the partnership of the Church with the agriculturists of the country, in a material portion of the produce of the soil is a subject deeply affecting the general interests, and in which information must be always of value.[39]

In a volume that intervenes between the book on the office of churchwarden and that on tithe, *A treatise on the laws of game, and inland fisheries in Ireland*, Finlay offers practical advice to future authors tempted to follow in his footsteps. 'It is not,' he argued,

38 Finlay, *Miscellanies*, p. 57.
39 *A treatise on the law of tithes*, preface, p. vi.

> by those massive publications, each of which seems in bulk like a small law library; which, oddly enough, are called by the authors, 'abridgements'; and might better bear the name of legal encyclopaedias, that law, in all or any of its branches, can be made distinct to the public or the profession. Such publications, however extended, can still be no more than an outline of the law; it is only by treatises on the subjects of law seriatim, that the knowledge of its rules and obligations can ever be made distinct and intelligible. The great Sir William Jones[40] was of this opinion, which he illustrated by the proposition that a map of the world however accurate, could never make a geographer; in fact, it must be a series of maps of particular countries, by which that knowledge can be communicated.[41]

Finlay employed the occasion of the self-same preliminary observations (as he calls them) again to stress the difference between the law in England and the law in Ireland. The uncritical, slavish adoption of the views of Blackstone was particularly highlighted. 'There is no branch of our law', Finlay wrote,

> in respect to which there is so much error abroad, as on the subject of the Game Laws. Our law books, and some of our best law books, abound with these errors; they in a great measure owe their origin to the celebrated Commentator on the laws of England; they have derived a sanction from the name of Blackstone; and the uncommon circulation of his immortal works has given diffusion and currency to his doctrines and mistakes. Amongst such mistakes those on the Game Laws are the most prominent, and they have been adopted by many law writers without examination and as authority.[42]

40 On Jones, now see D.J. Ibbetson, 'Sir William Jones, 1746–1794', in T.G. Watkin (ed.), *Legal Wales: its past, its future* (Welsh Legal History Society, vol. 1) (Cardiff, 2001), p. 41.

41 *A treatise on the laws of game and inland fisheries in Ireland* (1827), preliminary observations, p. xii.

42 Ibid., preliminary observations, p. iii.

The volume contains a number of surprises. In 1822, as his *Miscellanies* of 1835 reminds us, Finlay had addressed a series of letters to the Irish government on Ireland's internal resources. In this *mixum gatherum* of projects brought forward by Finlay – the development of canals, the development of paper manufacture, the exploitation of the country's granite, etc. – time and space were predictably devoted to inland fisheries.[43] An enthusiasm once aroused was unlikely to diminish, and it is no cause for wonder that in his treatise of 1827 Finlay quotes verbatim from a number of parliamentary debates, and supplies statistics on the amount of fish to be found in an inventory of the best-known of Ireland's angling waters. Finlay in his preliminary observations did not let slip the opportunity both to state the obvious and imply the need for preventative legal measures in support of this resource.[44] 'The extraordinary superabundance of fish in the inland waters of Ireland', he wrote,

> has been alleged at all times by the highest authorities, by Spencer, Temple, and Lord Bacon, Boate, Barton and Younge; and in modern times by Daniel, the great writer on this subject in England, as well Newenham, Tighe, Sampson and all the writers on the statistics of this country.[45]

The inclusion of a separate section on the game laws is certainly enlivened by Finlay's account of an actual criminal prosecution in which he had appeared as defence counsel. This was a case heard in the Head Police Office in Dublin in November 1826 when two men called Newman were successfully prosecuted under a statute of 1815 for shooting at game without a licence. The incidents had taken place north of Dublin at 'Finglas-wood' and 'Spring-mount'. What had been shot was

43 Finlay, *Miscellanies*, ch. 4.

44 Cumming stresses this point in his flyer. The volume, it is there stated, 'explains and sets forth the Statutes on the Laws of Inland Fisheries, against committing waste on the breed of Fish, and for encouraging that great source of national subsistence, which in Ireland is more abundant and obstructed, than perhaps in any other country.'

45 *A treatise on the laws of game, and inland fisheries in Ireland*, preliminary observations, p. xi.

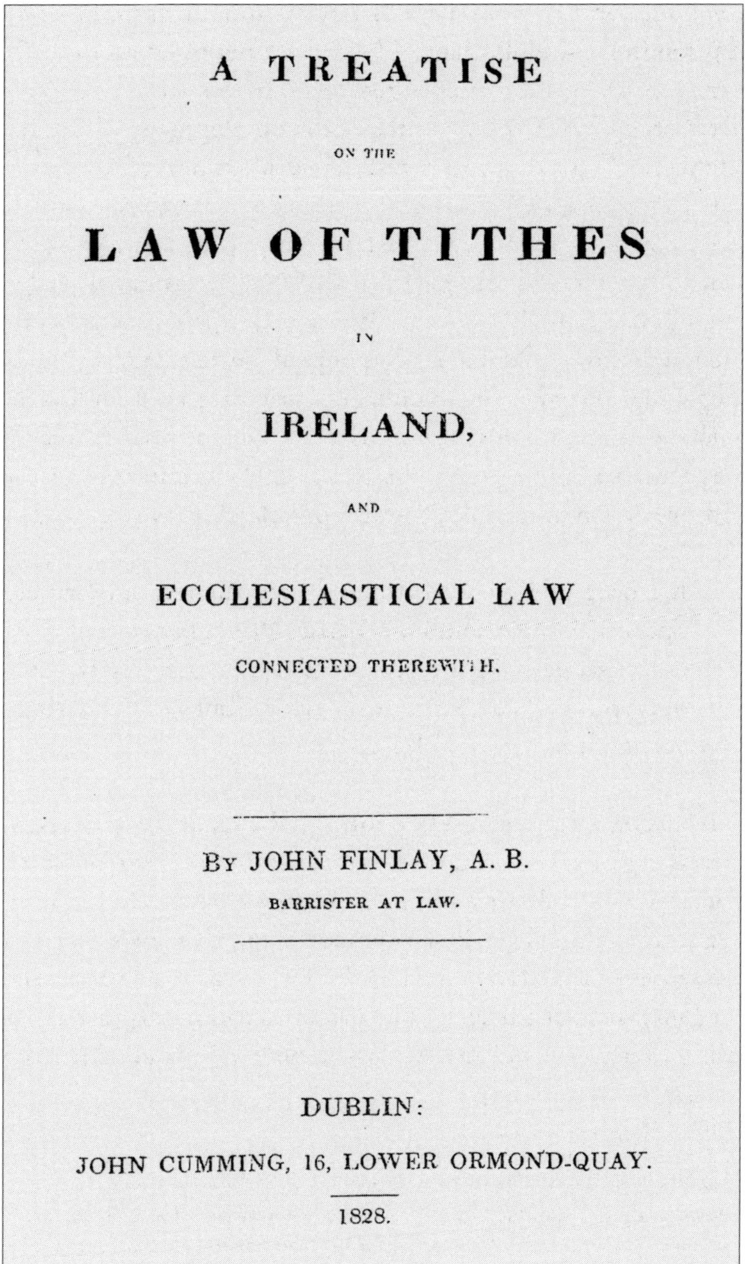

A TREATISE

ON THE

LAW OF TITHES

IN

IRELAND,

AND

ECCLESIASTICAL LAW

CONNECTED THEREWITH.

By JOHN FINLAY, A. B.

BARRISTER AT LAW.

DUBLIN:

JOHN CUMMING, 16, LOWER ORMOND-QUAY.

1828.

2 Title page of Finlay's excellent *Treatise on the law of tithes* (1828)

a landrail or corncrake. Finlay reproduces a press report of the proceedings, adding, however, that the newspaper had failed to discern the key question in the case.[46]

The lengthy title of Finlay's text on the law of renewals adequately conveys a sense of its content. Here the focus of attention was the Irish equity associated with leases for lives where there had been a failure to renew in timely fashion when the critical last life had fallen in. In what for Finlay was an unusual departure, the penning of a dedication – in this instance a dedication to Lord Plunket, at the time chief justice of the Irish Common Pleas – the opportunity was found for yet another comment on differences between Irish and English law, differences not always grasped in England where the House of Lords exercised final appellate jurisdiction since the enactment of the Acts of Union in 1800. A difficulty existed, Finlay wrote, where 'law peculiar to one country is referrible [*sic*] to a court of appeal in another country, where that law is imperfectly known'. Addressing Plunket directly, Finlay continued,

> Your peculiar and favourable position for this purpose, as an Irish lawyer and a British peer, to whom this peculiar equity is known, must give you advantages in making it intelligible on a tribunal to which it is not generally familiar, and afford you facilities at least to promote a uniformity of opinion on this subject for the future.

In the circumstances, he trusted that the volume 'may have some title' to Plunket's attention.[47]

The dedication to Plunket contained a separate passage where the purpose of introducing legal treatises to the profession and to interested members of the public was again to be addressed. Putative legal authors were once more brought within Finlay's sights. 'One of the benefits from law publications', Finlay bravely proclaimed,

46 Ibid., pp. 111–14. The statute in question was 55 Geo. III, c. 26.
47 Finlay, *A treatise on the law of renewals*, dedication.

> is, that by reminding courts of their previous views, our judicial authorities are preserved from self-contradiction or contrariety of decision, and enabled to maintain the uniformity of the law.[48]

That creed was certainly adhered to by an equally admirable volume dealing with virtually the same subject-matter, James Lyne's *A treatise on leases for lives*, brought out by the Dublin publishers Hodges and Smith in 1837.[49]

Cumming's flyer of August 1830 furnishes a detailed synopsis of the content of another of Finlay's legal texts, one, as it happens, to be succeeded and, in the end, superseded by books brought out by a long line of other authors. This is how Cumming sought to advertise Finlay's *The law of landlord and tenant for Ireland*, first published in 1825:

> A third of this Work is occupied with Law peculiar to this Country, and not to be found in any other book on Landlord and Tenant, such as Tenures and Rents, peculiar to Ireland – Leases for Lives renewable for ever, with the doctrine of their Renewals – Registration of Leases – Common – Estovers – Planting – Registration of Trees – a practical view of the Law of Tithe in Ireland, which, by a Statute, assumes the character of Rent, and takes that name – Ejectment by Civil Bill-Distress, with the Tenant's remedy where he had paid his rent to one Landlord, and is distrained by another for the same gale. Throughout the rest of the Work, where the Law is in general common to the two Countries, the numerous differences are carefully marked, where the English Treatise would mislead the inquirer. Subjects of general importance, but not much dwelt on by previous writers, have been amply discussed, as the Law concerning the off-going Crops, Lodgings, Emblements, Notices to Quit, Incorporeal Hereditament, etc. Finally, the general Law

48 Ibid. For same preliminary remarks on the handling, post-1800, by the United Kingdom House of Lords of Irish appeals, see Osborough, *Studies*, pp. 244–7. There it is related that Finlay's trust in Plunket as an honest broker in the Lords would seem to have been misplaced.

49 James Lyne, *A treatise on leases for lives, renewable for ever* (Dublin, 1837).

> affecting all descriptions of Estates, and the legal rights and duties of Tithe-owner and Tithe-payer, as well as Land-owner and Agriculturist, are arranged, and interspersed with the latest decisions.

Cumming's advertisement for Finlay's *Digest*, another pioneering achievement of the man, captures the essence of what for all students of modern Irish legal history remains an indispensable work of reference. 'From the Voluminous nature of Law Publications in England', the advertisement runs,

> Digests and Indexes to their Reports have become numerous and essential; indeed without them a collection of Law Books cannot be useful or available; yet, strange to say, no Index or Digest to the numerous reported cases on Irish Law has ever, until now, been offered to the Public. Indeed the number and dispersed nature of these Reports, together with the circumstance that so many of them had no Index to the particular Publication, was a great discouragement to an attempt of this kind.
>
> It is presumed that this Work must afford extraordinary facilities to both the Legal Professions, not only in Ireland, but in Great Britain, where many of our Reporters are received as authorities. And, further, that this Work must be generally useful to all persons interested in the Law of Ireland.

A second edition of Finlay's treatise on landlord and tenant and a third edition of that on churchwardens and other parish officers emerged from John Cumming's presses in 1835. The same year Finlay also saw through the press his *Miscellanies*, a collection of essays going back to the 1820s on a variety of topics. The volume also included the text of the address he delivered to Trinity College's Historical Society as a recent graduate and young barrister back in 1809. Some of this material had previously appeared in print.[50]

Finlay was to pen a second dedication – this time to the marquis of Anglesey – a dedication which, on account of the personal matters to

50 See publications listed in notes 10 and 18 above.

which it alludes, may strike the modern reader as needlessly obsequious, however much the sentiments expressed by Finlay will have been deeply felt. The dedication refers to Finlay's advancement to the bench which occurred when Anglesey was lord lieutenant. Finlay expresses his gratitude to Anglesey, who, in his distribution of the patronage of the crown, 'when pressed by high interests from different quarters, selected for promotion a man without such aids, and vindicated my labours from a long neglect'. 'This distinction', Finlay continued, 'has been to me a source of much benefit – a matter of great pride'.[51]

After 1835, Finlay appears to have rested on his laurels. There were to be no more books. His work as an assistant barrister presumably constituted the principal preoccupation. Still intimate with O'Connell, he made a new friend in Thomas Moore. They had dinner together in London in 1836,[52] and Finlay's wife was of the party when there was another dinner together in 1838.[53] It was on this last occasion that Moore recalled an observation of Finlay's that night on Irish history: 'The lies are bad, and the truth still worse'.[54] Back in Dublin, a Sunday in August 1841 found John Finlay and Thomas Moore walking together along the banks of the Liffey.[55]

A year or so after giving up his two remaining judicial offices, on 11 May 1856, John Finlay died at his home in Lower Cumberland Street. His funeral took place three days later at Glasnevin. The interment was to be recalled by W.J. Fitzpatrick many years later:

> Glasnevin Cemetery was thronged on May 14th, 1856, for on that day a once well-known man, John Finlay, LL.D., was buried in the presence of troops of his co-religionists, as well as of Catholics and Dissenters, who remembered how ungrudgingly he had given his great talents in furtherance of their emancipation.[56]

51 Finlay, *Miscellanies*, dedication.
52 Thomas Moore, *Memoirs, journal and correspondence*, ed. Lord John Russell, 8 vols (London, 1853–6), vii, 145.
53 Ibid., vii, p. 232.
54 Ibid.
55 Moore, *Memoirs*, vii, 299.
56 Fitzpatrick, *Dublin Catholic cemeteries*, p. 71.

Printing history

Editing and publishing historical documents in nineteenth-century Ireland

RAYMOND GILLESPIE

As print and book production grew dramatically in nineteenth-century Ireland historians contributed, perhaps, more than their fair share to that expansion in the world of print. Yet what most of those who practised the historian's craft produced is now little read. If a historical work of the nineteenth century is used today it is usually as part of a historiographical exercise or a curious example of a long dead methodology. There are, of course, significant exceptions to this generalization. Most of those nineteenth-century publications that are still consulted today have proved enduring not for their narrative or their analysis but for the documents that they reproduced. Nineteenth-century editions of crucial texts relating to the Irish past, however unsatisfactory they may be, are still in use today. In some cases this is a matter of convenience, but in others it is a necessity. The destruction of documents in the Public Record Office fire of 1922 has meant that nineteenth-century editions and calendars of, for example, the patent rolls or the deeds of Christ Church cathedral are all that remains of a once-extensive archive.

The potential created by cheap print and new publishing strategies in the nineteenth century for the publication of manuscript sources was exploited by a wide range of groups and individuals. Celtic scholars, in particular, saw the advantages of producing texts that could be used by both historians and philologists. The activities of some of this group are explored in this essay while others are dealt with elsewhere in this volume. The practice of editing and publishing texts in Ireland was not, of course, new. As early as the 1630s the Dublin antiquarian James Ware had published a selection of chronicles and annals which he considered important to understanding the Irish past.[1] In the eighteenth century Walter Harris, a relative of Ware by marriage, and John Lodge gathered together another selection of documents that they considered would throw light mainly on the seventeenth century.[2] For a different set of reasons older works, such as John Temple's *History of the Irish rebellion* (London, 1646), were reprinted during the eighteenth century because they were held to contain political messages that were still of relevance. However, what is striking is how limited this activity was in the eighteenth century. While scholars such as John Lyon reorganized the archives of the major private institutions in Ireland, including Trinity College, Dublin, and Christ Church and St Patrick's cathedrals, and produced finding aids for them, none of these made their way into a wider world through the medium of print. As a consequence throughout the nineteenth century Lyon's masterpiece the 'Registrum novum', which transcribed the medieval deeds of Christ Church, lay unnoticed in the cathedral archives from the time of its creation in the 1740s.[3]

In the nineteenth century all that was to change as a result of two factors. The first was the change in publishing and marketing that allowed the production of editions of texts at lower costs than before as technology improved and markets expanded. The second was the growth in the number of groups and individuals who were interested, for a variety of reasons, in making archival sources more widely available. These

1 James Ware (ed.), *The historie of Ireland collected by three learned authors* (Dublin, 1633).

2 [John Lodge (ed.)], *Desiderata curiosa Hibernica: or, A collection of state papers* (Dublin, 1772); Walter Harris (ed.), *Hibernica or some antient pieces relating to Ireland* 2 vols (Dublin, 1747–50).

3 Raymond Gillespie, 'The archives of Christ Church cathedral' in *Irish Archives,* 5:2 (Winter 1998), pp. 5–6.

groups, of which the most important were the various agencies of central government, provided the infrastructure which allowed a generation of talented individuals, such as Sir J.T. Gilbert, to engage in the process of producing editions of texts for the use, primarily, of historians. Central government may have been the major player in this process but it was not the only one. Local government, in the shape of Dublin Corporation, was responsible for Gilbert's monumental calendar of that city's records which appeared from 1889 onwards. Moreover, enthusiastic learned societies from the Irish Archaeological Society in the 1840s through the Ossianic Society in the 1850s to the Irish Texts Society established in the 1890s all published editions of texts to suit the scholarly predilections of their members.[4] From the 1850s local learned journals, such as the *Ulster Journal of Archaeology* and the *Journal of the Kilkenny and South East Ireland Archaeological Society*, carried editions of documents in their pages as a way of illustrating curiosities from the past. A few brave individuals, such as Gilbert, risked their own funds to publish editions of works that they thought important, in his case on the 1640s.

While central government was undoubtedly the most significant agency in making documents available in print in the nineteenth century a category as wide as this conceals as much as it reveals. Within this broad category a wide range of groups were at work and a number of philosophical constructs as to the best way of presenting documents to the user coexisted. The most fundamental dilemma was how best to represent on the printed page the content of the manuscript and hence make it available in a useable form to the reader, who may have been some distance from the original document and hence unable to check it. For some the answer was to print the entire text in a verbatim transcript, while others advocated the idea of a calendar, or abstract, of particular classes of documents. Essentially, such calendars were the translation into print of finding aids, or working notes, prepared by those who had custody of such documents. As such they were necessarily selective, reflecting administrative necessity and the constraints of the making of

4 For these societies see Damien Murray (ed.), *Romanticism, nationalism and Irish antiquarian societies, 1840–80* (Maynooth, 2000); Pádraig Ó Riain (ed.), *Irish Texts Society: the first hundred years* (London, 1998).

the calendar rather than faithfully representing the complete content of the document. Principles of selection, for instance, were often unclear, if set out at all. Moreover, for material in remote locations, such as the Public Record Office in London, historians in Ireland came to rely on printed calendars rather than consulting the original documents. As a result, for most of the nineteenth, and indeed the twentieth century, the questions that historians asked were shaped by the nature of the published calendars to which they came to have ready access through the medium of print.

In the Irish context the use of the printed calendar was pioneered by the Irish Record Commission of 1810–30. The Commission represents the first determined attempt to grapple with the problem of providing access to the Irish public records which were un-centralized and, for the most part, unorganized and hence inaccessible. As an offshoot of the English Record Commission the Irish Commission tended to follow practices established in England through the work of the Record Commission there. Before the establishment of the Irish Commission in 1810 English Commissioners had inspected the Irish records and found both the state of preservation and publication sadly wanting.[5] The Irish Commission quickly realized how different the Irish situation was to that of England. The secretary of the Irish commissioners, William Shaw Mason, reported to the commissioners in 1811 that the English strategy of publishing what they regarded as the most important original records in government archives was not appropriate in Ireland since 'from the mutilated condition, and in many cases the total want of Calendars and Indexes, even in the most antient and extensive repositories, it becomes quite unnecessary for the present, to consider of any selection from such imperfect materials [for publication]'. That did not mean that publication was undesirable and Shaw Mason identified a number of classes of records, such as the statute rolls as well as material in Trinity College, Dublin, that might be published at a later date.[6] For this reason the Irish

5 The Irish Commission is dealt with in M.C. Griffith, 'The Irish Record Commission, 1810–30', *Irish Historical Studies*, 7:25 (March, 1950), 17–38.

6 *First report to His Majesty's commissioners on the public records of Ireland* (Dublin, 1811), pp. 33–9.

Record Commission concentrated on the making of manuscript calendars and indexes, in order to be able to find documents, rather than publishing complete texts. The impact of the commissioners in Ireland was limited, mainly as a result of disorganization within the Commission and a lack of familiarity with the texts with which they were working. Indeed the lack of technical skill in calendaring documents and then preparing the documents, or the calendars, for publication was a problem that Shaw Mason had foreseen in his first report to the Record Commissioners, urging the commissioners to give it their 'earliest attention and investigation'.[7] That they failed to do so would have long-term ramifications. Despite this the commissioners did manage to publish some of their calendars. In 1827 the calendar of the chancery inquisitions for Leinster was printed and the following year a calendar of the patent rolls from Henry II to Henry VII was published. However when set against the number of publications of the English Record Commission over the same period (between forty and fifty) this seemed poor value for the £100,000 expended in Ireland. The appearance of another calendar of chancery inquisitions for Ulster in 1829 did not impress the authorities and the Irish Record Commission was wound up in 1830, leaving a calendar of the patent rolls of James I partly printed. This was completed after the Commission's demise.

The termination of the work of the Irish Record Commission in 1830 effectively put an end to attempts by central government within Ireland to publish historical texts and calendars relating to Ireland. Perhaps most importantly the lack of a central authority in Irish record keeping, especially a Public Record Office which was not established until 1867, ensured that no one had either access to documents or the resources to push forward such a policy. However developments in England were to revitalize these attempts in the late 1850s and 1860s. Unlike Ireland the momentum of publication established by the English Record Commission, and its successors, was continued by the Public Record Office following its establishment in 1838, some thirty years before the foundation of its Irish counterpart.

7 *First report to the commissioners*, p. 39.

While the Irish Record Commission had made little impact on the world of record publishing, in the form of either texts or calendars, before its dissolution in 1830 it had gone some way towards creating a more centralized form of record keeping. By the 1840s the Customs House in Dublin came to have that function for Exchequer records. In 1848 commissioners were again appointed to inquire into the state of the public records in Ireland and there was some discussion of seconding English officials from the Public Record Office there to reform the Irish system of record keeping.[8] This interest in Irish affairs by those in the London Public Record Office continued into the early 1850s as Rowley Lascalles' *Liber munernum publicorum Hiberniae*, which had been made for the Irish Record Commission in the 1820s but remained unpublished, was prepared for publication by F.C. Thomas, the secretary of the English Record Office. Some of his interest may have been personal since Rowley's son William was then employed by the English Public Record Office.[9] All this activity seemed to be moving inexorably to the foundation of an Irish Public Record Office on the English model, as the report of the 1848 commissioners recommended and as eventually occurred in 1867. An integral part of that process was the creation of a publication policy. The rationale for organizing, cataloguing and centralizing records was to make them widely available to those who visited the repository and the logical outcome of that was to provide guides and texts in print for those who could not do so for reasons of distance. The emergence of record offices and record publication were part of one process. Inevitably this was not a neutral process, even more so given the scale on which print could now circulate ideas and texts. Dealing with the past, and record publication was no exception, was a highly charged political and cultural activity in the nineteenth century. Political conditions in the 1860s meant that the matter of publication was one to which the London Treasury was turning its thoughts and it resulted in reports on the Carte and Carew manuscripts being made, according to the Master of the Rolls, to show that 'Her Majesty's government have not been unmindful of the wishes expressed by the Irish nation to have

8 John D. Cantwell, *The Public Record Office, 1838–1958* (London, 1991), pp. 114–15.
9 Cantwell, *Public Record Office*, pp. 161–2

the materials for Irish history collected and that they have made an attempt to discover and disclose important documents connected with the historical literature of Ireland'.[10]

In this context the Irish chancery clerk James Morrin was commissioned to prepare calendars of what might well be regarded as the most useful of the Irish records: the chancery rolls. In the first volume, published in 1861, Morrin declared that his aim was to 'rescue some part of the most important of our national muniments from the comparative oblivion and obscurity which, by reason of the difficulty of access and labour of research, they now lie involved'. In short, Morrin took the view that such volumes would facilitate the 'research of persons engaged in historical investigations and enquiry' such as antiquarians, historians or legal practitioners.[11] In the companion volume, published the following year, Morrin seemed to row back somewhat, stressing the practical use of his volume as 'a storehouse of facts and documents for the use of the statesman, the lawyer and the antiquary, nor will they be found, I trust, unworthy the regard of the scholar and the historian'.[12] For Morrin, his work in making the contents of the patent rolls available was of such evident importance that he could not understand the shortage of other calendars and provided extensive historical introductions to those he edited. That was to prove his undoing. After the publication of the second volume of his calendar of the patent rolls Morrin's work came under a fierce attack in 1863 from an 'Irish archivist', a pseudonym for none other than John T. Gilbert. In a published pamphlet, Gilbert openly accused Morrin of plagiarism in the material used in the historical introductions to his calendars, mistakes in his calendaring and incompetence in reading the manuscripts on which he worked. Gilbert also raised the wider question of the state of record keeping in Ireland. The matter was even brought to the attention of parliament with the result that the Treasury established an inquiry which broadly exonerated Morrin, a fact

10 [T.D. Hardy and William Brewer], *Report to the Right Honourable master of the Rolls on the Carte and Carew Papers* (London, 1864), p. 3.

11 James Morrin, *Calendar of patent and close rolls of chancery in Ireland, Henry VIII to 18th Elizabeth* (Dublin, 1861), p. xliii.

12 James Morrin, *Calendar of the patent and close rolls of chancery in Ireland from the 18th to the 45th of Queen Elizabeth* (Dublin, 1862), p. lxxviii.

1 John T. Gilbert

which raised Gilbert's ire further.[13] One further volume of Morrin's calendar of patent rolls, that for the early part of the reign of Charles I, appeared in 1864 while the final volume was left incomplete and never published.[14]

There is little doubt that even by the standards of their own day Morrin's calendars left a good deal to be desired. Certainly when compared with the Record Commissioners' calendar of the patent rolls of James I and J.C. Erck's two-volume 'repertory' of the same material made in the 1840s, Morrin's work was perfunctory. But, in some ways the attack on Morrin was a sideshow to the main point that Gilbert wished to make. His concern was not with the usefulness of the volumes but the

13 [J.T. Gilbert], *On the history, position and treatment of the public records of Ireland* (2nd edn, London, 1864); [J.T. Gilbert], *English commissioners and Irish records* (London, 1865); Cantwell, *Public Record Office*, pp. 227–9.

14 James Morrin, *Calendar of the patent and close rolls of chancery in Ireland: Charles I years 1 to 8* (Dublin, 1864); The National Archives: Public Record Office, London [hereafter TNA], T/1/653B.

more general issue of the shoddy treatment of Irish archival material in contrast to the way in which archives were dealt with elsewhere. Thus the fact that these less-than-satisfactory calendars had the stamp of official authority was, to the nationalist-minded Gilbert, simply another example of how second-rate work that would not be acceptable in England was deemed to be adequate in Ireland.[15] He later applied this principle to physical production of editions as well as to editorial practice. In a letter pasted into the presentation copy of the first volume of Gilbert's *Calendar of the ancient records of Dublin* (Dublin, 1889) in the library of the Public Record Office in London, Gilbert noted that this Irish product in 'the mechanical parts – printing, binding and illustrations' was as good as anything that had appeared from the English Public Record Office. This argument – that Ireland was not a provincial backwater and needed to be treated almost better than England – Gilbert would frequently use later when, for example, he felt that Irish materials (mainly edited by him) were not being produced quickly enough in the Rolls Series, then in financial difficulties. There is no doubt but that the publication of archival texts, especially Morrin's calendars, became a politically contested issue in the 1860s, reflecting, as some saw it, the treatment of Ireland in general. It certainly hastened the establishment of the Irish Public Record Office in 1867 and the exploration of the Carte and Carew papers as a placatory gesture to Irish nationalists in the early 1860s.

The debacle surrounding the publication of Morrin's calendars effectively put an end to any schemes for extensive record publication by the Irish administration. That does not mean that the idea of publishing official records was abandoned. In particular the deputy keeper of the public records, a post created in 1867 with the new Irish Public Record Office, was, like his English counterpart deeply committed to the policy of making archival sources available in print. The first Irish deputy keeper, Sir Samuel Ferguson, was, by his own admission, a practical man whose own account of his years as deputy keeper emphasized the col-

15 [Gilbert], *History, position and treatment of the public records*, p. 141 [Gilbert], *English commissioners and Irish records*, pp. 1–2.

2 Sir Samuel Ferguson

lection, arrangement and making available of records. However as Ferguson's deputy, J.J. Digges La Touche, recalled 'it was Dr Ferguson's most ardent desire to continue the work of the Record publication which had been begun so many years before by the Irish Record Commissioners ... [this] was a consummation very dear to him'.[16] However the problems of record publication that the experience with Morrin's calendars had thrown up, together with increasing Treasury parsimony, obliged Ferguson to look for alternative outlets for record publication. Ferguson, like his English counterpart, began to experiment with the publication of calendars of the Record Office's collections as appendices to the annual reports of the Deputy Keeper. The process was initiated in 1875 with the publication of a calendar of the fiants for the reign of Henry VIII and in 1888 the calendar of Christ Church deeds began to appear. The first part, comprising 467 entries, was a calendar of the deeds transcribed into the 'Registrum Novum'. This was followed by two other instalments covering the years up to 1602 and an index was provided in 1895. Later

16 Lady Ferguson, *Sir Samuel Ferguson in the Ireland of his day* 2 vols (Edinburgh, 1896), ii, pp. 332, 336–7.

calendars of the Irish pipe rolls would also appear in the same format. Through this stratagem a range of calendars, mainly of medieval records, found their way into the public domain.

If the printed calendar of records failed to develop as a significant publishing venture in nineteenth-century Ireland it was perfected at the Public Record Office in London. The deputy keeper of the records there after 1861, Thomas Duffus Hardy, was deeply committed to the idea of making records available to the 'inquirer living at a distance' through publication, a theme that runs through his reports as deputy keeper.[17] However, the question remained as to how that was to be achieved. In the 1830s the London State Paper Office had favoured the publication of complete texts in preference to calendars. From the 1830s it had been preparing and producing volumes of the correspondence of the reign of Henry VIII, the Irish material appearing as parts two and three. In 1854 when the State Paper Office and the Public Record Office were merged the practice of publishing transcripts of state papers was discontinued and replaced by the idea of publishing calendars of this material, almost accidentally developed by the Record Commission in the 1830s.

From the mid-1850s the Public Record Office began preparing the calendars of one of their largest collections relating to Ireland, the state papers. In 1860 the first Irish volume, dealing with the period 1509–73, was published. The format of the volumes had already been determined in calendars of the domestic series of state papers which had begun appearing in 1856. The editor of the first five volumes of Irish state papers, Hans Hamilton, had no Irish connections. He came to the Public Record Office from the English State Paper Office and he remained on the editorial side of the Record Office until his retirement in 1888 on health grounds.[18] This came just at the point when the London Public Record Office began expanding its calendaring activity, especially under the deputy keepership of Henry Maxwell Lyte after 1886. It became clear that the Record Office could not manage this expanded role within its own resources. The practice of drawing Irish scholars into cataloguing activity can be identified from the middle of the 1860s. In the 1870s H.S.

17 Cantwell, *Public Record Office, 1838–1958*, pp. 215–16.
18 Ibid., pp. 169, 238, 288, 316, 572.

Sweetman, an Irish barrister who was a regular user of the search room in the Public Record Office, was employed to extract Irish medieval material from English records up to 1485 to be published as separate volumes. The experience was not a happy one though, and the project remained unfinished at Sweetman's death.[19] Again when J.S. Brewer, a Public Record Office official, was appointed to calendar the Carew papers he was to be assisted by the Irish Protestant William Bullen.[20] However such collaborations could be fraught with difficulty. The original scheme of appointing the impressive scholars J.H. Todd and William Reeves to prepare calendars of the Irish state papers had to be abandoned since both were Protestants and this was perceived as a difficulty. A confessional balance was deemed necessary and this requirement produced one of the most formidable cataloguing duos of the nineteenth century: the lawyer and expert on the 1650s, J.P Prendergast, and the president and former professor of ecclesiastical history in St Patrick's College, Maynooth, C.W. Russell. While J.T. Gilbert, Morrin's nemesis, was considered for the post he was not approached. Russell and Prendergast had first worked together preparing a report on the Carte material in the Bodleian Library in Oxford.[21] The result had been a printed overview of the collection and although Russell seems to have begun work on a calendar of the whole collection in the format of the calendars of state papers it was never finished.[22] However, in 1869 Russell and Prendergast were commissioned to calendar the Irish state papers of the reign of James I to the standards established for the other state papers calendars. The first volume appeared in 1872 and two more in 1874. By the end of 1876 parts of the final volume of James I's reign were complete but in 1877 Russell was badly injured in an accident, from which he never really recovered, and the final volume of the calendar was completed by Prendergast alone in 1880 after Russell's death.[23] What is significant about the state papers project is that the resulting calendars were intended not

19 Ibid., pp. 261–2, 300. 20 Ibid., p. 226.

21 For Prendergast's reaction to the appointment of Russell see Cantwell, *Public Record Office*, p. 227.

22 The incomplete catalogue is now in the Russell Library, Maynooth College.

23 Ambrose Macaulay, *Dr Russell of Maynooth* (London, 1983), pp. 254–62.

only to be scholarly ornaments but rather were to be used to make material on the Irish past accessible. The Treasury, for example, sanctioned review copies for the *Irish Times*, *Freeman's Journal* and the *Dublin Evening Mail* and this linked with an expanded Stationery Office operation in Ireland to ensure that copies were easily available.[24] Copies were also distributed as gifts to the libraries of major institutions. The remainder of the Irish state papers to 1680 were calendared between 1900 and 1912, after a gap of some twenty years in the project, by R.P. Mahaffy, the eldest son of J.P Mahaffy, the Hellenic scholar and provost of Trinity College, Dublin. R.P. Mahaffy was a Cambridge graduate and a lawyer working in London with little historical experience. His principal recommendation for the post was the perception of his Irishness since the resumption of the state papers project after twenty years was, no doubt, influenced by the fact that in 1899 an acrimonious debate had taken place in the House of Commons about access to the Irish state papers. Some nationalists had argued that they had been excluded from access to the state papers while others, such as W.E.H. Lecky, had been given preferential access.[25] As so often in the nineteenth century, political considerations shaped the form and timing of official record publications.

While the published calendar became synonymous with the publication of national archives from the London Public Record Office in the nineteenth century that body did not monopolize the technique. In 1869 Queen Victoria appointed a Royal Commission on Historical Manuscripts. Its remit, as with so many attempts at record publication in the nineteenth century, was to make available calendars of collections of private papers in great houses following on from inspections. There were two features of this Commission. First, from its origin the publication of reports, or calendars, on collections, as opposed simply to their compilation, was part of the brief of the Commission. The first report appeared in 1870. Secondly, Ireland was placed clearly within the commissioners brief. Indeed, the original commissioners included a number

24 TNA, PRO/1/42, PRO letterbook, letters of 17 July, 19 July, 28 July, 30 July, 9 Aug., 14 Aug., 1877; Hugh Barty-King, *Her Majesty's Stationery Office: the story of the first 200 years, 1786–1986* (London, 1986), pp. 22, 30–48.

25 Cantwell, *Public Record Office*, pp. 344–5.

of people with Irish credentials, including Lord Talbot de Malahide, Charles William Russell, president of St Patrick's College, Maynooth (and maker of the calendars of the Irish state papers) and the antiquarian Charles Graves, bishop of Limerick.[26] The appointment of the Irish inspector was something of a foregone conclusion. Whatever one thought of the character of J.T. Gilbert after his spat with James Morrin he was both industrious and knowledgeable on Irish records. Queen Victoria had brought the commission into existence on 2 April 1869 and by 4 October of the same year Gilbert had inspected the records of Kilkenny, Limerick, Waterford and Cork corporations. He also proposed to make a calendar of Lord Ross's manuscripts for the forthcoming report. By October of the following year he claimed to have prepared a report on Lord Granard's manuscripts.[27] Gilbert urged himself on, viewing the reports of the commission as an important way of demonstrating the significance of Irish records, and particularly those of the 1640s, to a wider world and using his position to obtain resources for the publication of records. Thus, commenting on the Ormond collection in Kilkenny Castle on which he was working in the 1870s he observed to T.D. Hardy in the Public Record Office in London 'I do not know any work that would be more gratifying to the Irish nation than a calendar of this collection which has never yet been opened to the public and respecting which there is the greatest curiosity'.[28]

In practice, Gilbert drove himself too hard and adopted a creative approach towards his expenses. The result was that he almost bankrupted the commission's Irish operation in the 1880s and 1890s when funds were tight. He refused, for instance, to adhere to demands to shorten his reports and to reproduce fewer documents in full.[29] As on other occasions he resorted to what can only be described as blackmailing the commission by playing the Irish card so that they would provide more funds for his activities. As he stated unambiguously to the secretary of the com-

26 For a list of Commission members, see Royal Commission on Historical Manuscripts, *Manuscripts and men* (London, 1969), p. 84.
27 TNA, HMC/1/159, 4 Oct. 1869 Gilbert to Brett; 5 Oct. 1870, Gilbert to Brett.
28 TNA, HMC/1/162, 2 Nov. 1871, Gilbert to Hardy.
29 TNA, HMC/1/163, 12 Jan. 1889, Cartwright to Gilbert.

mission 'from what I have heard the House of Commons could not object to an increased allocation for the commission if it were specially for publishing their Irish reports'.[30] It is true that he generated a large range of reports, many of which were popular. He noted in 1877 that 'the Dublin agents for the sale of government publications had an unusually large demand for copies of the appendix to the tenth report containing the accounts of the Irish collections. They tell me they never before sold so many copies of any of the publications of the commission'.[31] As against this he left a large number of reports unfinished at his death, some of which were completed later, and he himself suffered two breakdowns.[32] He also became fixated by earlier collections, especially the papers of the marquis of Ormond, which resulted in disproportinate resources being devoted to them. However, without Gilbert the business of record publication, especially through the Historical Manuscripts Commission calendars, would have been a less significant feature of Irish scholarly life in the nineteenth century. Gilbert, perhaps more than anyone else, expanded the limits of the system ensuring Irish records got favourable treatment and the quality of calendaring was of a high standard, in some places being nearly complete transcripts of texts.

While the idea of the published calendar may have dominated the thinking of the practical men in archives during the nineteenth century it did not sweep all before it. There were those who saw the calendar as a useful tool but no substitute for the publication of complete texts. Through the late eighteenth and into the nineteenth century the scholarly approach involved in the publication of complete texts had been most associated with those who wanted to collect evidence on the origin and development of their own nations. State building gave rise to monumental editions of texts, sometimes built on earlier ways of organizing similar projects such as those of the Bollandists or the Maurists in the case of the church, which were intended to vindicate national

30 TNA, HMC/1/164, 13 Dec. 1893, Gilbert to Cartwright.

31 TNA, HMC/1/163, 2 July 1887, Gilbert to Cartwright.

32 For an overview of his career see Toby Barnard, 'Sir John Gilbert and Irish historiography' in Mary Clark, Yvonne Desmond and Nodhlaig P. Hardiman (eds), *Sir John T. Gilbert, 1829–1898* (Dublin, 1999), esp. pp. 100–7.

arrangements or aspirations. In Germany the *Monumenta Germaniae Historica* or in England the Rolls Series can be seen as examples of such developments.[33] Ireland was not wanting in this sort of activity. The Revd Charles O'Conor's *Rerum Hibernicarum scriptores* privately published between 1814 and 1826, for instance, conforms to this sort of activity in a more modest way than some of the larger European schemes. O'Conor, as librarian to the marquess of Buckingham at Stowe, had unique access to a range of early Irish historical sources, including important sets of annals, which he edited in four volumes. Few, however, shared either O'Conor's access to key primary texts or the financial resources necessary for their publication. The more usual way of ensuring that texts that were regarded as central to an understanding of a particular facet of the past were published was through the formation of a learned society. In England, for instance, local record publication grew dramatically in the nineteenth century as county or other societies issued volumes of special interest to their members. In Ireland the focus for such groups was not so local. The lack of a substantial county gentry is certainly one factor in this but another is the rise of the Dublin middle class with the resources necessary to subscribe to such works.

Throughout the nineteenth century a series of scholarly societies grew up to meet the demand in Dublin and elsewhere for scholarly erudition. These societies viewed the creation of printed artefacts as a means of demonstrating one's cultural and political preferences. Three groups were of particular importance in this endeavour; the Irish Archaeological Society (founded in 1840), the Celtic Society (founded in 1845) and the Ossianic Society (founded in 1853). The first two of these merged in 1854 to form the Irish Archaeological and Celtic Society which continued to function until 1873. The Ossianic Society concentrated mainly on medieval and modern Irish texts while the others published a mixture of medieval Irish language texts and Latin or English documents. Each society had its own emphasis. The Ossianic Society tended to draw its supporters from the nationalist population while the other groups, drawn more from the upper echelons of Irish society, looked to Celtic Ireland

33 For these schemes see, David Knowles, *Great historical enterprises* (London, 1962).

as a golden age that could unite Protestantism and Irishness. Most significantly, a cadre from Trinity College provided access to the Dublin University Press which undertook their publications.[34] By the end of the nineteenth century more specialist societies began to venture into the world of publishing historical documents. In the 1890s the Irish Texts Society was formed initially to produce an Irish dictionary but by 1899 it had issued its first work, comprising an edition of three tales by Douglas Hyde.[35] At another level the Royal Society of Antiquaries of Ireland also turned its attention to the printed word as an 'antiquity' publishing a series of editions of texts, four of which appeared in the 1890s. Little is known of motivation and method here but the hand of James Mills from the Public Record Office of Ireland can be clearly discerned in a number of their publications. A study of this group and their activities is long overdue.[36]

In addition to the local learned societies within Ireland the publication of complete editions of texts relating to Ireland had one further outlet: through the Rolls Series attached to the Public Record Office in London. The Rolls Series began publication in 1858 and almost as soon as the first volume had appeared some in Ireland began to consider what advantage was to be obtained for them since the series's remit was the publication of editions relating to both Great Britain and Ireland. As early as March 1857 the Dublin antiquarian John Dalton was demanding the inclusion of Irish material in the series. Probably about the time the first volume appeared, J.H. Todd, librarian of Trinity College and a distinguished Celtic scholar, became involved. Todd was well used to fund raising from London and in 1836 he had been on a committee appointed by the Royal Irish Academy to solicit money from the chancellor of the exchequer for the publication of the Annals of the Four Masters. As soon as he was aware of the Rolls Series he wrote to John

34 For a fuller history of these societies, see Murray, *Romanticism, nationalism and Irish antiquarian societies*.

35 For the Society see P. Ó Riain (ed.), *Irish Texts Society*, esp. pp. 2–35.

36 For an introductory overview, see Aideen Ireland, 'The Royal Society of Antiquaries of Ireland, 1849–1900', *Journal of the Royal Society of Antiquaries of Ireland*, 112 (1982), 72–92.

Romilly, the master of the rolls and an activist on behalf of the Public Record Office, inquiring what it intended to do to 'represent the history of the Celtic inhabitants of the United Kingdom' in the series. This approach was not entirely uninterested. There was a more pressing matter. The Irish Archaeological Society, of which Todd was a leading member, was in financial difficulties and he seems to have hoped that it might be possible for the two bodies jointly to publish an edition of the Annals of Ulster, a transcript of which had been made.[37] In the event the project did not materialize but what did appear as a result of this initiative was Todd's edition of the *War of the Gaedhil with the Gaill* in 1867. He did not have the honour of the first Irish text in the Rolls Series since that belongs to William Hennessy's edition of the *Chronicon Scotorum* which had appeared the previous year. However Todd's intervention in the 1850s was of some importance since it established clearly the principles for editing Irish texts for the Rolls Series. First, while Latin texts in the series were left untranslated, Irish language texts had to be issued in parallel form with English translations on facing pages. This followed the style adopted by the Irish Archaeological Society from the 1840s. Even if it could not be fully understood by most readers the inclusion of the Irish text was deemed essential although it gave rise to considerable technical problems for the printer attempting to align the Irish text and the translation. Secondly, the Irish text needed its own special typeface, and could not be set in Roman type. Todd urged that the Rolls Series should print their Irish works in Dublin since 'the Irish type used in London is so bad being taken from very incorrect matrices'. For perfection the type designed by Petrie for editions of Irish manuscripts should be used.[38] Thus, as Todd insisted, Irish texts required special treatment and needed to look Irish as well as being Irish. Even a text as arcane as the *War of the Gaedhil with the Gaill* could be said to be 'selling very

37 TNA, PRO 37/64, 4 Mar. 1858, Todd to Romilly.

38 Todd's letters on the production of the volume are all in the TNA, PRO 37/64. For this type and its cultural milieu, see Dermot McGuinne, *Irish type design: a history of printing types in the Irish character* (Dublin, 1992), pp. 99–117; see also pp. 125–6 below.

quickly in this country [Ireland]' when it appeared in 1867, presumably at least as much because of its appearance as a genuine Irish work as for any other reason.[39]

Almost as soon as the earliest Irish language texts were issued in the Rolls Series the editors were giving some thought to the political implications of their inclusion. After the publication of the *Chronicon Scotorum* an internal Public Record Office memorandum noted, 'it would be advisable if possible to allow the Irish to have another work in the government series as it is not improbable that application will be made to the Treasury to have a specific portion of our parliamentary grant allocated to Ireland direct'.[40] No such allocation was made but a number of Irish texts were published in this series. All were done to the standard prescribed by Todd for the earliest of them. Thus Irish texts were not simply made available but came with a message about the importance of language and manuscript in the Irish past.

The second tradition of Irish texts, those medieval works in Latin, were less exotic and more akin to the Latin texts from England that made up most of the series. Thus they could not be used as effectively as a statement of Irishness. For some, such as J.T. Gilbert, the inability to edit manuscripts in the Irish language text series was a severe embarrassment and in 1865 it was revealed that he tried to publish an edition of a text by the late Eugene O'Curry as his own. As the subsequent Public Record Office investigation put it, Gilbert was trying to pass off 'another man's labours as his own' which was held to be 'questionable' and the edition was suppressed.[41] Despite this, one of the earliest texts worked on for the Rolls Series was James Graves's edition of the Latin roll of the proceedings of the king's council in Ireland 1392–3, the original manuscript of which was in Kilkenny Castle. Although Graves began work on this in 1859 it did not appear until 1877 because of illness.[42] Undaunted by his failure to publish an edition of an Irish language text in the Rolls series Gilbert was forced into the tradition of editing Latin

39 TNA, PRO 37/42, 7 Jan. 1867, William Hennessy to T.D. Hardy.
40 TNA, PRO 37/42, 'Memorandum on publication of *Chronicon Scotorum*, 1864'.
41 TNA, PRO 37/31 'Memorandum on Mr Gilbert's proposal'.
42 Graves file, TNA, PRO 37/33.

manuscripts. Prefiguring his calendar of the records of Dublin, on which he had produced a report for Dublin corporation in 1866, in 1867 he began his *Historic and municipal documents, Ireland* which appeared in 1870 but which seems to have been substantially complete by the middle of 1869. This was subjected to a savage review in the *Athenaeum* to which he felt the need to compose a reply to the editors of the Rolls Series.[43] Gilbert, however, was not prepared to rest on his laurels, however unstable, and in the 1880s proposed to the Rolls Series an edition of the Credi Mihi, which later appeared privately, and editions of the medieval Registers of St Thomas's and St Mary's in Dublin. By this stage the Rolls Series was experiencing financial difficulties and the Public Record Office was under pressure to cut back on new commissions. Gilbert and others interpreted this as yet another slight on Irish records and, after correspondence in *The Times* and questions in the House of Commons by John Dillon and T.P. O'Connor, Gilbert's volumes were sanctioned by the Treasury.[44] It is hard to miss the veiled threats in Gilbert's letters of the 1890s to J.J. Cartwright, secretary of the Historical Manuscripts Commission that although the Rolls series may have been closed for England 'their lordships of the Treasury may not be disinclined to sanction some special volumes for Ireland to meet the wishes of the public in that matter'.[45] Record publication had a power greater than scholarly influence in achieving its aims.

In historical terms the most enduring contribution of the print revolution of the nineteenth century was the making available of texts and the opening up of collections of documents for scholars. Through the prefaces of editions and calendars runs the leitmotif of the editor opening the inaccessible, especially to those living at a distance from archives, or the previously unorganized. In this the expansion of print was a crucial factor. Other developments, such as the organization of central archives and developments in the study of diplomatic and palaeography, were no doubt important but without the mechanization of print which

43 All in the Gilbert file in TNA, PRO 37/31 and HMC/1/159/ 5 July 1869, Gilbert to Brett.

44 Knowles, *Great historical enterprises*, pp. 126–7.

45 TNA, HMC/1/163, Gilbert to Cartwright, 21 Nov. 1891.

reduced production costs and extended runs, much of this would have remained the preserve of the few. More than anything else the calendar was the way in which the agents of this change, principally central government, gave effect to their programme to disseminate information about the national past. The Public Record Office and the Historical Manuscripts Commission used calendars to revolutionary effect, to the extent that they replaced the original documents in the minds of some. However, more specialized societies continued in an older tradition of issuing edited texts of what they regarded as crucial documents to their interested members. In this way printed documents about the past moved rapidly into the public sphere of nineteenth-century Ireland and became part of a wider debate about Ireland, Irishness and Irish history. Access to the past became embroiled in the political concerns of the present. At one level this had to do with what the documents contained. At the beginning of the nineteenth century Sir William Betham had opposed the publication of inquisitions by the Record Commission because 'such printing may do much harm by exciting the discontents of the descendants (whether real or pretended) of the attainted proprietors' and so 'irritate the public mind'.[46] Later, Gilbert would observe of the difficulties associated with Irish historical manuscripts 'I may mention that annual celebrations of events of those times 1690–91 are still kept up with ... so you will see that the ground is dangerous and great caution is necessary'.[47] However all was not that crude. The debate between Gilbert and Morrin over standards of editing revolved around other issues of archives and their use as symbols of national culture, civility and pride. The advent of cheap print and the consequent rise of record publication certainly had the power to edify, educate and inform but by spreading ideas it also had the power to make subtle mischief on a scale not hitherto seen.

46 Griffith, 'Irish Record Commission', p. 26.
47 TNA, HMC/1/160, 30 May 1877, Gilbert to Cartwright.

William Reeves and the edition of the *Life of St Columba*

The Continental connection

JEAN-MICHEL PICARD

THE CURRENT PRESSURE ON academic publishers to produce best-sellers is a development of the late twentieth century. At the time of the nineteenth-century industrial revolution, high volume of sales became the aim of newspaper owners, but mass production was never the primary aim of the publishers of scholarly books. For the most part, their interest lay in recognition by their peers and by the community of scholars of the quality of their books. This tradition and code of values can be traced back to the sixteenth century in the publishing houses of Amerbach in Basle, Manutius in Venice, Bade and Estienne in Paris or Gryphe in Lyon and was maintained in subsequent generations. Nowadays saints' lives and books on hagiography, such as those published by Four Courts Press or Brepols, are unlikely candidates for widespread interest, but these publications (especially the *Legenda Aurea* by Jacobus de Voragine, both in Latin and in translations) had provided a small but steady income for European printers in the early modern era.

The publication of scholarly editions – based on research, with introductions and critical commentary – was difficult to achieve and was usually undertaken by learned societies, who also raised funds to cover the high cost of production. In the seventeenth century, societies such as the Bollandists in Belgium or the Société de St-Germain-des-Près in Paris fulfilled this function and undertook the task of publishing new collections of *Acta sanctorum*, using scientific and critical methods.[1] The Société de St-Germain-des-Près included outstanding scholars like D'Achéry, Du Cange, Baluze, Lamy, Lenain de Tillemont, and the Maurist Jean Mabillon, but their work was not without difficulty. The latter met opposition within his own Benedictine order from members who believed monks should devote themselves to prayers and manual labour rather than to study. He also was *persona non grata* in Rome because of his scientific approach and criticism of the popular veneration of saints whose existence could not be attested. His scholarly work could not have continued without the protection and financial support of King Louis XIV and his minister, Colbert.

In contrast, in the 1830s, Jacques Paul Migne started to use, for religious publications, the same strategies as the large commercial publishers. His idea was to make available works of hagiography and patristics at a low price and in high numbers so that all members of the clergy could afford to buy them. His printing house, the Imprimerie Catholique, employed more than 300 workmen and between 1844 and 1855, he produced the 217 volumes of the *Patrologia Latina*. The 247 volumes of the *Patrologia Graeca* were published between 1856 and 1866. The *Patrologia* was a great achievement in terms of volume and circulation. However, because of the speed of production, the editions of texts were unsatisfactory and contain large numbers of errors. Migne's enterprise ended sadly: his business was affected by a fire at the printing works in 1868 and by the Franco-Prussian war of 1870. In order to finance his operations, he started trafficking in religious art and in stipends for Masses. As a result, he was suspended from his priestly functions by the archbishop of Paris, condemned by Rome in 1875 and died penniless in the same year.

While the publication of the *Patrologia Latina* was in progress on the continent, the Irishman the Reverend William Reeves was earning

1 See E. De Broglie, *Mabillon et la Société de Saint-Germain-des-Près au XVIIIe siècle*, 2 vols (Paris, 1888).

1 Revd Willam Reeeves, scholar and bishop of Down and Connor

European fame for the limited and model publication of *The life of St Columba*, founder of Iona.[2] Between his graduation from Trinity College, Dublin in 1835 and his election to the Royal Irish Academy in 1846, Reeves had acquired a solid reputation in Ireland for thorough scholarship. His *Ecclesiastical antiquities of Down, Connor and Dromore*, published in 1848 by Hodges and Smith, had confirmed this reputation. The book was reviewed in the *Dublin University Magazine* by Samuel Ferguson, who wrote:

> Among the most signal contributions which have been made to the Irish Library since the revival of historical learning among

2 *The life of St Columba, founder of Hy; written by Adamnan, ninth abbot of that monastery. The text from a manuscript of the eighth century; with the various readings of six other manuscripts, preserved in different parts of Europe. To which are added copious notes and dissertations illustrative of the Columban institutions of Ireland and Scotland* – By William Reeves, D.D., M.R.I.A., curate of Kilconriola in the Diocese of Connor – Printed at the University Press by M.H. Gill for the Irish Archaeological and Celtic Society – Dublin – 1857.

> us, we assign an eminent place to this work of Mr Reeves. When it is considered how much has been accomplished in this way within the last five years, Mr Reeves' friends will perceive how highly we estimate his services. In that high esteem, we do him justice. He has earned his place with Petrie and O'Donovan. Like them, he has earned it by grave, erudite and patriotic labour: may he long enjoy the eminence so honourably and usefully achieved.[3]

Acclaimed in 1857 as 'a work of the highest historical importance' and as 'the most valuable contribution ever made to the history of the early Celtic Church',[4] his edition of the *Life of St Columba* was the culmination of many years of work.

The project had started in 1848. On 19 June of that year, Reeves wrote a long letter to John O'Donovan which included queries about Adomnán's *Vita Columbae*. O'Donovan's reply reassured him on the genuine value of the *Life*.[5] The following year, Reeves was ready to advance his project further. On 16 June 1849, he sent a letter to the famous Scottish antiquarian Joseph Robertson, then secretary of the Maitland Club in Glasgow, offering to prepare an edition of the *Life* for publication by the Maitland Club.

> I have long wished to see that unique specimen of ancient ecclesiastical biography placed before the antiquarian public in an accessible and inviting form. [...] It is very desirable, for the antiquarian credit of both Scotland and Ireland, that a new edition, in the form of a monograph, should appear, illustrated with

3 *Dublin University Magazine* 31 (1848), 207.

4 Lady Mary Ferguson, *Life of the Right Rev. William Reeves, D.D.* (Dublin, 1893) pp. 47–8.

5 Correspondence of John O'Donovan, University College Dublin, Special Collections, MS JO'D L33: '8 Newcomen Place, North Strand, Dublin. July 14th 1848. [...] I have no doubt about the *Vita Columbae* having been written by Adamnan. He mentions his predecessors in the abbacy of Iona and he mentions himself *mihi Adamnano*. His description of the plague satisfies me [...]'. I am grateful to the UCD Library staff in Special Collections, who facilitated my study of this correspondence.

the topographical and historical matter which recent researches in the two countries have brought to light. By a happy concurrence of circumstances St Columba may be regarded by the Scotch and Irish as common property: he was born in Ireland, but he died in Scotland; he was ordained in Ireland, but his ministrations chiefly lay in Scotland; he was educated in Ireland, but he taught in Scotland; while as regards his biographer, the same may be said, and both are held in equal veneration on either side of the Channel. [...]

In this manner there would be presented to the Scottish antiquary a work free from all speculations, and giving the earliest information about the most venerable institution which his country can boast of.

Robertson replied immediately:

51, Windsor Terrace, Glasgow,
23rd June 1849.

Reverend Dear Sir,
I had the pleasure of your note of the 16th, on Wednesday morning. I laid your very interesting outline of the edition of Adamnan's *Life of St Columba* before a meeting of the council of the Maitland Club on Thursday afternoon; and I am glad to inform you that the proposal was received with universal favour, with more general and hearty satisfaction, indeed, than I remember to have been accorded to the first mention of any new undertaking.

The Council instructed me to thank you for the trouble you had taken, and to ask if you could make a rough estimate of the probable extent of the work, and the probable expense of transcripts, collating, and editing. I rejoice in the hope that my last act as secretary of the Maitland Club will probably secure for that society the distinction of a work so desirable as the *Life of St Columba*, edited with the diligence, learning, and accuracy

> which are so conspicuous in the *Ecclesiastical antiquities of Down, Connor, and Dromore.*

In spite of this positive answer, the project did not proceed immediately. Reeves was unable to provide the estimate required by Robertson, mainly because the cost of gathering together the sources on the continent could not be established in advance.

Reeves had stressed the Irish and Scottish setting of the *Life of St Columba*, but omitted to mention that the majority of medieval manuscripts of Adomnán's work were kept in continental libraries, including the oldest one, written before 713 by Dorbbéne, head of the Iona school. This precious manuscript was sent from Iona to Northern France during the eighth century and then to the upper Rhine valley, where its final destination was the monastery library in Reichenau on Lake Constance.[6] It was rediscovered there in 1621 by the Irish Jesuit Stephen White, who copied it. This copy was given to Colgan, who printed it as *Vita quarta S. Columbae* in his *Trias thaumaturga* (1647). Sometime before 1639, White had lent his copy to Ussher, who made an abstract of it.[7] John Tyrell, grandson of Ussher still owned this copy in 1662. Another copy was used by the Bollandist Francis Baert, for inclusion in the second June volume of the *Acta sanctorum* (1698).[8] Reeves had access to these printed texts and obtained a loan of the Ussher manuscript from its then owner, John C. Nicholl of Merthyr Mawr (Mid Glamorgan, Wales). He had also recently made an important discovery in learning, from his friend and mentor James Henthorn Todd, of the whereabouts of the original manuscript written by Dorbbéne. Todd had received some fac-

6 See Jean-Michel Picard, 'Schaffhausen Generalia I and the textual transmission of Adomnán's *Vita Columbae* on the continent', in P. Ní Chatháin & M. Richter (eds), *Ireland and Europe in the early Middle Ages: texts and transmission* (Dublin, 2001) pp. 95–102.

7 James Ussher, *Britannicarum ecclesiarum antiquitates* (Dublin, 1639), reprinted in C.R. Elrington, J.H. Todd and W. Reeves (eds) *The whole works of the Most Reverend James Ussher*, 17 vols (Dublin, 1847–64) vi, p. 236.

8 Johannes Colganus (ed.), *Triadis thaumaturgae seu divorum Patricii, Columbae et Brigidae, trium veteris et maioris Scotiae, seu Hiberniae, Sanctorum Insulae communium patronorum acta* (Louvain, 1647) pp. 336–72; Franciscus Baertius (ed.), *Acta sanctorum*, Junii II (Antwerpen, 1698) pp. 197–236.

simile pages of this manuscript from a fellow antiquarian in Zurich, Dr Ferdinand Keller, who had re-discovered the manuscript in the Municipal Library in Schaffhausen.[9]

Ferdinand Keller is the famous archaeologist who excavated the site of Obermeilen on Lake Zurich in 1854, thus establishing the beginnings of wetland archaeology. He also published in 1868 the report of the La Tène excavations, well known among specialists of Celtic archaeology. Following the discovery of two tumuli near Zurich in 1832, Keller had founded the Society of Antiquaries of Zurich (*Antiquarische Gesellschaft in Zürich*) and its journal, the *Mitteilungen der Antiquarischen Gesellschaft in Zürich*. In 1845, he had come across the Dorbbéne codex in Schaffhausen and taken the manuscript with him to Zurich for examination. Facsimiles were made, which he later sent to J.H. Todd for further identification.

Correspondence now in the Royal Irish Academy shows that Reeves contacted Keller directly in autumn 1850. Keller replied with a long memorandum including a good description of the manuscript, its key characteristics (*Adomnanus*, never *Adamnanus*; *Insula Iova*, never *Iona*; diacritic signs on dative and ablative plurals and on proper nouns) and a history of the initial rediscovery of the manuscript at Schaffhausen in the eighteenth century by the Benedictine historian Moritz Hohenbaum Van der Meer (†1795), prior and keeper of the monastic archives at Rheinau. He also offered to have the manuscript transcribed for Reeves at a cost of £2 10*s.* 0*d.* and a half pounds sterling.[10] Attached to the memorandum were several documents which Keller had obtained from his contacts in Schaffhausen and in Rheinau: (a) a note in German from the historian Johann Jakob Mezger of Schaffhausen, dated 2 January 1851, introducing a copy of the notice written by Hohenbaum Van der Meer when he examined the Dorbbéne codex, *Notae in manuscriptum*

9 The date and circumstances of the transfer of the Dorbbéne manuscript from Reichenau to Schaffhausen are not known. See René Specht, 'Wie kam Dorbénes Abschrift von Adamnáns *Vita Sancti Columbae* in die Stadtbibliothek Schaffhausen?', *Schaffhauser Beiträge zur Geschichte* 65 (1988), 103–9.

10 This memo and attached documents are now preserved in the Library of the Royal Irish Academy [hereafter RIA], Dublin, MS 12 W 29. I wish to thank the librarian, Siobhán Fitzpatrick, and her staff for facilitating my research.

Scaffusianum, inscriptum à tergo Vita Columbani. The notice clearly identifies *Columbanus* as Colum Cille (†597) and the names and dates of Adomnán (†704) and Dorbbéne (†713). It also identifies the Schaffhausen manuscript as the codex copied by Stephen White at Reichenau in the seventeenth century; (b) a note in German from Father Fridolin Waltenspül, prior of Rheinau (dated 8 January 1851) concerning the work of Moritz Hohenbaum Van der Meer and (c) a third note in German from Father Leodegar Ineichen, Librarian at Rheinau – he would become the last abbot of Rheinau in 1859 – quoting from the relevant passage of the *miscellanea* written by Van der Meer.

Reeves did not know German and never attempted to learn it, as he confesses in a letter written to his cousin, and future wife, Charlotte Townley on 13 August 1891: 'It is an incalculable loss to me that I don't know a word of German, for the line of my studies has a great deal to do with the works of German scholars, who are now deep in the investigation of Irish, both language and antiquities'.[11] However, Keller wrote in English and translated the German documents in his letter. All the information concerning the Schaffhausen manuscript in the introduction to the *Life of St Columba* is derived from Keller's memorandum, including the quotation in German attributed by Reeves to Van der Meer, which is in fact the extract written by Father Leodegar.[12]

Between the end of 1850 and the end of 1853, Reeves maintained a regular correspondence with Keller, who supplied the data needed for the edition of the *Vita Columbae*. Keller again borrowed the Dorbbéne manuscript from the library in Schaffhausen to have it copied in Zurich. He secured the services of a teacher of Greek and Latin, Joseph Conrad Morf, to carry out the transcription. This copy was sent to Reeves on 22 January 1851, with two bills in German from Morf dated 10 and 18 January 1851 amounting to £4 10*s*. 0*d*. sterling. Keller's accompanying letter explains why the price exceeded the £2 10*s*. 0*d*. sterling previously agreed: two collations had to be made because, at the first attempt, Morf had standardized the spellings instead of reproducing the original Hiberno-Latin spelling of Dorbbéne and Keller asked him to do it again.

11 Ferguson, *Life of William Reeves*, p. 173.
12 Reeves, *Life of St Columba*, p. xv and notes.

Morf reckoned he had earned less than one shilling per hour and his helper (*famulus*) only 3 or 4 pence on account of the difficulty in reading the Irish script. In order to keep costs down, Keller had to do the facsimiles himself. Receipt of these documents made it possible for Reeves to start work on his principal manuscript.

The next hurdle was the collation of another important manuscript in the history of the textual transmission of the *Vita Columbae*. In the ninth century, a monk of St Gall compiled an abridged version of Adomnán's text, which became the most widely known version in continental Europe.[13] This was the version printed by Canisius in 1604 and later by Migne in his *Patrologia Latina*.[14] The original manuscript was still in the library of the monastery at St Gall (MS 555) and Reeves wanted to include it in his apparatus. Keller had examined it in St Gall and warned Reeves that it was a copy of the Dorbbéne codex and that it might not be of any value for a critical edition. However, Keller offered the services of a friend, Professor Wyss, to do the transcription of the manuscript. Reeves does not appear to have received a transcription from Wyss and had to wait until July 1853 to get readings of MS 555, mainly because of the rivalry between Ferdinand Keller and Karl Johann Greith, dean of the cathedral at St Gall. Both men had done research on the work of Irish monks in Switzerland and each saw himself as the only expert in that field. Furthermore, they belonged to different parties in the Catholic Church, Greith being an Ultramontane and Keller a liberal. Greith finally sent the facsimiles of the first page of MS 555 and of the Carolingian drawing of St Columba,[15] through the medium of the Zurich city librarian Johann Jakob Horner, who also advanced the money due to the draftsman. Horner, who was the nephew of the celebrated astronomer and mathematician Johann Kaspar Horner, was a friend of Keller's and was also one of J.H. Todd's antiquarian correspondents. In

13 See Jean-Michel Picard, 'Adomnán's *Vita Columbae* and the cult of Colum Cille in continental Europe', *Proceedings of the Royal Irish Academy* [hereafter *PRIA*] Sect. C, 98 (1998), 1–23.

14 Heinrich Canisius, *Antiquae lectiones*, 6 vols (Ingolstadt, 1601–4) v, part 2, pp. 559–621; J.-P. Migne, *Patrologia Latina*, 221 vols (1844–64) lxxxviii, col. 725–76.

15 Both facsimiles are reproduced in the *Life of St Columba*, facing p. xxvi and facing p. xxviii.

his correspondence with Reeves, Keller does not hide his annoyance with Greith's involvement: 'Mr Horner has sent the 15 francs de France (which the artist had the impudence to charge or Greith the audacity to propose) for these two drawings'.[16] As we shall see below, Reeves did not take sides in this rivalry and made use of both Greith and Keller in order to obtain reliable readings for St Gall MS 555.

In the summer of 1853, Reeves became interested in a third continental manuscript, Munich Clm 6341, formerly belonging to Freising but probably written in St Gall in the last third of the ninth century.[17] This time, the transactions were easier. He wrote to the library and received a reply from Dr Conrad Hofmann offering to have a full transcription done for £2 sterling. This letter was in German and Reeves had it translated by his friend Robert Mac Adam, the Belfast antiquarian (dated 2 August 1853). Conrad Hofmann did the work quickly and sent his transcription on 16 October 1853, with a facsimile of the beginning and the end the manuscript done by Herr Schenk of the Munich Library for a fee of ten shillings sterling.

In her biography of Bishop Reeves, Lady Mary Ferguson states that the work on the *Life of St Columba* 'was reserved to be a solace to Reeves in his day of sorrow and hours of solitude, and was accomplished at the close of 1856'.[18] In fact, most of the work was finished before the death of his first wife, Emma Reeves, on 12 October 1855. In 1853, the Irish Archaeological Society (founded in 1840) and the Celtic Society (founded in 1845) had merged to form the Irish Archaeological and Celtic Society, with as its patron Prince Albert and as its president the duke of Leinster. The secretaries were J.H. Todd and John T. Gilbert. Members of the council included Eugene O'Curry, Charles Graves, John O'Donovan and George Petrie. The first large project of the society was the publication of the *Leabhar Imuinn* edited by J.H. Todd.[19]

The second major project was Reeves' *Life of St Columba*. The deci-

16 RIA, MS 12 W 29.

17 Natalia Daniel, *Handschriften des zehnten Jahrhunderts aus der Freisinger Dombibliothek* (Munich, 1973) pp. 58–61.

18 Ferguson, *Life of William Reeves*, p. 47.

19 J.H. Todd, *Leabhar Imuinn. The book of hymns of the ancient Church of Ireland* (Dublin, 1855).

sion to print Reeves' work is reported by Aquilla Smith, medical doctor and antiquarian, in a letter written on 6 April 1855 to his friend John Windele in Cork:

> Last week the council made an arrangement to put to press immediately the *Life of St Columba* edited by the Rev. William Reeves, a work on which he has been long engaged and the entire [work] is ready for the printer. Dr Reeves is energetic and this volume will be ready for delivery before the close of this year.[20]

As Smith was also the honorary treasurer of the Irish Archaeological and Celtic Society, he was well aware of the financial implications of such an undertaking. The society was financed by subscription from its members, which consisted of £3 admission and £1 per annum for full members, and £1 per annum and no admission fee for associate members. The average annual income of the society in each of the three years 1855 to 1857 was only £185. The society could have had more members, but at the time of the merger in 1853, the Celtic Society had split and a large number of its members had left to form the rival Ossianic Society.[21] The costs of production of the *Life of St Columba* are itemized in the Report of the Auditors of the society for the year 1857:

23 July 1856: to Messrs Hanhart for lithographic fac-similes for Life of St Columba	£21
17 Feb 1857: to M.H. Gill on account of printing the Life of St Columba	£350

This sum clearly exceeded the annual income of the society in any given year and in order to make publication possible, members were asked to wait for two years to allow the costs to be spread ('1855 and 1856: *The life of St Columba* by Adamnan, ninth abbot of Hy [or Iona]. The two parts

20 RIA, MS 4 B 15/30.
21 Robert Somerville-Woodward, 'The Ossianic Society, 1853–1863', *Pages* 6 (1999), available at www.ucd.ie/pages/99/articles/somewood.html.

are bound in one volume for the convenience of members') and the Bannatyne club of Edinburgh was asked to contribute £135 for 100 copies which would be printed on their own paper.[22] With the printing of the *Leabhar Imuinn*, edited by Todd, and Reeves' *Life of St Columba*, the Society had spent more than £500 in less than three years. It was not bankrupt, but from a healthy balance in credit of £138, which had earned interest in 1854 and 1855, it was down to nil by the end of February 1857.[23]

It is noteworthy that compared to the expenditure on printing and publication, the costs involved in collating the manuscripts were relatively small. At the beginning of the project in 1849, in the letter quoted above, Joseph Robertson had enquired about probable costs connected with obtaining transcripts, but he need not have worried. Reeves never travelled to the continent and the total cost of gathering the continental material was less than £8. This was the equivalent of about half a year's wages for a ploughman and two weeks of Reeves' own annual income, but is a small sum compared to the total cost of publication at £371. The book was prepared for printing at the Dublin University Press by M.H. Gill. As usual, there were author's corrections and late additions, including the description and readings from a forgotten British Museum manuscript, recently restored by Sir Frederick Madden in 1852, and examined by Reeves *in extremis*, in October 1856.[24] Reeves finally completed his introduction on 25 November 1856, with a word of compliment for Gill and the Dublin University Press, which 'had fully maintained its character for elegance'.[25] Each copy was personalized, with the name of the individual subscriber printed in red on the recto of the title page: 'This copy was printed for [name], Member of the Society'. It was still at the binder's in March 1857,[26] but when it finally came out, it

22 *Report of the Irish Archaeological and Celtic Society* (Dublin, 1857).

23 Ibid.

24 This is British Library MS Cotton Tiberius D. iii (s.xii). See Reeves, *Life of St Columba*, pp. xxvii-xxxi and 456–8 and letter written from London to John O'Donovan, UCD, Special Collections, JO'D L 56, dated 25 Oct. 1856.

25 Reeves, *Life of St Columba*, p. xxxvii.

26 Letter from Reeves to Revd Alexander Irwin, precentor of Armagh and secretary to Primate Beresford, dated 4 March 1857: 'My "Columba" is in the binder's hand and will be *publici juris* in a week or so'.

received instant approval from all. The honours bestowed on Reeves in the years following the publication of the *Life* are testimony to the high esteem in which he was held, both in Ireland and Scotland: honorary membership of Society of Antiquaries of Scotland, limited to twenty-five members; the Cunningham Gold Medal from the Royal Irish Academy and an honorary LLD from the University of Edinburgh.

On the Continent, the main agent for the fame of Reeves' book was Charles Forbes René de Tryon, comte de Montalembert. Montalembert was an important figure on the French political scene in the 1850s. With his friends Félicité de Lamennais and Henri Lacordaire, OP, he had founded the liberal Catholic newspaper *L'Avenir* in 1830. This newspaper was highly polemical and was the voice of Ultramontane opinion in France, opposing the concordat between Church and State signed in 1801. Montalembert was a brilliant orator in the French parliament, advocating the cause of Catholic education and defending the freedom of the Church in a Republican state. He also was a prolific author and had been elected to the Académie Française in 1851. He was born in London in 1810 to an aristocratic father, émigré during the French Revolution and a Scottish mother who had converted to Catholicism and he was a fluent English speaker. Montalembert was a friend of Ireland and of the 'Irish' cause, which he vigorously promoted both in *L'Avenir* and in his speeches in parliament. He was a great admirer of O'Connell and had visited Ireland in September and October 1830 to meet him. On his way to Rome in February 1847, O'Connell had stopped in Paris, where he was officially welcomed by Montalembert. In the same year, he also took charge of the co-ordination of the fund raising organized in France for the relief of the Irish Famine. He kept a regular correspondence with friends in Ireland, mostly members of the ascendancy with whom he shared common interests in politics but also in antiquarian and historical matters. Among these were Edwin Wyndham-Quin, third earl of Dunraven and his brother-in-law William Monsell, Liberal MP for Limerick.

Dunraven was a member of the Irish Archaeological and Celtic Society and, as such, one of the subscribers of Reeves' *Life of St Columba*. He was aware of Montalembert's keen interest in saints' lives. The latter had published in 1836 a *Life of Saint Elizabeth of Hungary*, which had become a best seller and was reprinted nine times between 1836 and 1861.[27] He had also embarked on the major project of writing a history of western monasticism spanning six centuries from St Benedict to St Bernard. The final product would be a seven-volume work, *Les Moines d'Occident depuis Saint Benoît jusqu'à Saint Bernard*, published between 1860 and 1877, volumes 6 and 7 being edited by Aurélien de Courson after Montalembert's death in 1870. *Les Moines d'Occident* was quickly translated into other European languages and became a main reference book for Church historians. In England, the first volume was published in 1861, only one year after its French publication, by William Blackwood and Sons.[28] Following the publication of *The monks of the West*, Dunraven sent a copy of the *Life of St Columba* to Montalembert. He was acquainted with Reeves' previous work, but this latest book came to him as a revelation. The letter he sent Reeves is worth quoting in full:

La Roche en Breny, 15 January 1863

Reverend Sir,

I should be guilty of great injustice and great ingratitude if I refrained from expressing to you the extreme gratification I owe to the perusal or rather to the attentive study of your recent publications on the ecclesiastical and monastic antiquities of Ireland. I allude to the different tracts which you did me the honour of forwarding to me last year, and amongst which those on 'Early Irish Calligraphy', on the 'Ancient Abbatial Succession', and on 'Marianus Scotus', have afforded me much valuable information. But I feel still more obliged to you, as all lovers and students of Church history ought to feel, after having studied your

27 Montalembert, *Histoire de sainte Elisabeth de Hongrie, duchesse de Thuringe* (Paris, 1836).

28 *The monks of the West, from St Benedict to St Bernard*, by the Count of Montalembert, member of the French Academy, authorized translation, 7 vols (William Blackwood and Sons, Edinburgh and London 1861–77).

admirable edition of Adamnan's *Life of St Columba*, the gift of which I owe to my dear and excellent friend Lord Dunraven. For more than five and-twenty years, I have been more or less conversant with old and new books on hagiography, and I think I may safely say I have never met with a more satisfactory and more conscientious labour in that line than yours. I have been living for the past two months in constant communication with you, owing to Adamnan and St Columba. I have thought it my duty to mention your works in a very cursory view of Irish monasticism, which I have just been correcting, for the second edition of a work called *Les Moines d'Occident*. But I shall make a point of quoting you as often as I can in the third volume of this same work, which will begin by an account of St Columbkille's mission in the Scottish Isles. As soon as the second edition of the two first volumes is ready, I shall ask your leave to present them to you, as, from all that I have read of you, I feel sure that, although belonging to a different Church, you will not scorn the toil of a sincere Christian, even if he be styled a Jesuit or an Ultramontane fanatic by the sapient Great Britons who enlighten the world in the columns of *The Times* or *The Morning Post*. I have the honour to remain, Reverend Sir,

With sincere respect
Your obliged and obedient servant,
C. Cte. De Montalembert.

Montalembert was true to his word. Reeves' work is thoroughly acknowledged and praised in the third volume of *Les Moines d'Occident*, finally published in 1866 and translated into English for publication by W. Blackwood in 1867.[29] For example:

29 Montalembert dedicated this volume to his friend Edwin Wyndham Quin, earl of Dunraven, with an allusion to their trip to Iona in 1862 (*comiti itineris comissimo / amico in adversis probatissimo*). The section on Columba was quickly reprinted by William Blackwood and Sons as a popular booklet without any notes under the title *Saint Columba, apostle of Caledonia*, by the Count de Montalembert of the French Academy (Edinburgh and London, 1868).

> Adamnan's work [...] has just been reprinted, after a MS. of the eighth century, by the Rev. Dr. W. Reeves, for the Celtic Archaeological Society in Dublin, with maps, glossary, and appendix, Dublin 1857. This excellent publication, which is distinguished by an impartiality too rare among learned English authors, has rendered a considerable service both to the hagiography and to the national history of Ireland and Scotland.[30]

Other recent works by Reeves are also acknowledged: 'The ancient abbatial succession in Ireland' (1858), 'Memoir of the Church of S. Duilech' (1859), *The Culdees of the British Islands*, published by M.H. Gill in 1864.[31] In the same book, he also refers to Reeves and O'Curry as 'the two princes of Irish erudition'.[32]

Apart from its widely recognized scholarly value, the book was an important monument to the spirit of conciliation between Catholics and Protestants. The 1850s were a period of strong rivalry between both sides not only in Ireland, with the founding or revival of militant Protestant clubs on the one hand and, on the other hand, the hardening of Catholic positions around the question of the Papal States, but in Europe in general. Reeves was not unaware of the political situation and from the beginning of the project in 1849, he was concerned to pres-

30 *The monks of the West*, iii, p. 98, note 2.

31 Ibid. p. 283, note 1: 'Reeves, Memory of the Church of St Duilech, 1859'; cf. 'Memoir of the Church of S. Duilech', *PRIA* 7 (1862), 141–7; the paper had been read before the RIA on 11 April 1859 and published in the same year as a separate pamphlet, Dublin 1859; ibid., p. 288, note 1: 'Dr Reeves has thoroughly examined this curious question in a special paper, On the Ancient Abbatial Succession in Ireland, ap. Proceedings of the Royal Irish Academy, vol. vii., 1857'; cf. 'The Ancient Abbatial Succession in Ireland', *PRIA* 6 (1858), 447–51; read before the RIA 12 Jan. 1857; ibid., p. 294, note 1: 'Since the preceding note was written, a new publication by Dr Reeves, *The Culdees of the British Islands as they appear in History, with an appendix of evidences* (Dublin 1864), has summed up and ended all controversy upon this long-disputed question, and given the last blow to the chimeras of sectarian erudition'; cf. 'A memoir of the Céli-Dé, or Culdees, of Ireland and Great Britain', *Transactions of the Royal Irish Academy* 24 (1864) pp. 119–264, separately issued and printed by M.H. Gill under the title *The Culdees of the British Islands, as they appear in History, with an appendix of evidences*, Dublin 1864.

32 Ibid., p. 294, note 1.

THE

LIFE OF ST. COLUMBA,

FOUNDER OF HY;

WRITTEN BY ADAMNAN,

NINTH ABBOT OF THAT MONASTERY.

THE TEXT PRINTED FROM A MANUSCRIPT OF THE EIGHTH CENTURY; WITH THE VARIOUS READINGS OF SIX OTHER MANUSCRIPTS PRESERVED IN DIFFERENT PARTS OF EUROPE.

TO WHICH ARE ADDED,

COPIOUS NOTES AND DISSERTATIONS,

ILLUSTRATIVE OF THE EARLY HISTORY OF THE COLUMBIAN INSTITUTIONS IN IRELAND AND SCOTLAND.

BY

WILLIAM REEVES, D. D., M. R. I. A.,

CURATE OF KILCONRIOLA, IN THE DIOCESE OF CONNOR.

DUBLIN:

Printed at the University Press,

FOR THE IRISH ARCHÆOLOGICAL AND CELTIC SOCIETY.

1857.

2 Title page of *The life of St Columba*

ent a balanced view of early Irish ecclesiastical history, free from all prejudices. Montalembert, champion of the Catholic cause in France, alludes to these conflicts when he mentions to Reeves the attacks against him in the British press. In this respect, it is quite interesting to compare his high opinion of Reeves with his view of British writers in general:

> It is impossible to support, by facts of authentic documents, those doubts as to the orthodoxy of the Irish, which have been borrowed from the unsatisfactory and partial learning of English writers of the past century by various authors of our own day…[33]

and further

> Modern research has finally dispersed all the imaginary chimeras, as certain English and German writers who attributed these differences to a pretended influence of Eastern Christianity on the British Churches, of which no authentic trace exists; or more readily still, to a traditional repugnance on the part of the Celtic population to the yoke of Rome – a repugnance belied by the history of the past, as well as by the living testimony of the races, the most tenacious and most illustrious members of which, the Irish and the Armoricans have purchased, at the cost of the most generous and cruel sacrifices, the right of placing themselves in the foremost rank of the faithful children of the Church of Rome.[34]

Even J.H. Todd does not meet with his full approval:

> A learned Englishman of our own day, Dr Todd, in his *Monograph on St Patrick*, published on 1863, acknowledges that the Irish Church of the sixth century differed in nothing as to doctrine from the rest of the Catholic Church; but at the same time he maintains her independence of the Holy See.[35]

33 Ibid., p. 85.

34 Ibid., pp. 378–9.

35 Ibid., p. 85, note 1.

Obviously, one could agree on many points amongst scholars and gentlemen, but Rome would remain a delicate point for a long time to come.

Similar discreet allusions were also made by his Swiss correspondents. In December 1851, Ferdinand Keller sent to Reeves a letter which introduced the readings from St Gall MS 555 made by Karl Johann Greith, dean of St Gall:

> Dear Sir,
> This moment I receive the above notices of the Columba MS at S. Gall. The collation is, as you see not finished, but this fragment will enable you to judge of the value of the MS. Though I have not applied to Dean Greith, but to Mr Gmür for the collation, the former would not let escape an opportunity of showing his learning and his zeal in the catholic cause. I designedly have not mentioned your name to Mr. Gmür, but simply stated that some Irish clergymen were desirous of having the collation of that MS – leaving it uncertain whether they were Catholics or members of the Anglican Church. Mr. Greith being one of the most fiery Ultramontane priests of Switzerland would probably not have been very expeditious, had he known that a heretic had undertaken to publish the Life of a saint, so full of marvellous stories. In future we shall no more want the assistance of Greith, as I have found a person at S. Gall who will do such a conferring and copying business most promptly and correctly. I shall thank in your and your countrymen's names Dean Greith for these notices and send him some literary stuff that will be of use to him.
> Zurich, 9 December 1851[36]

In 1851, the memory of the civil war between the Protestant majority and the group of Catholic cantons (*Sonderbund*) which had seceded from the Confederation was still recent (1845–7). The political and military defeat of the *Sonderbund* resulted in the expulsion of the Jesuit order

36 RIA, MS 12 W 29.

from Switzerland and the domination of radical and liberal leaders in the new federal state constituted in 1848. However the Catholics soon reorganized, and the 1850s were a period of active campaigning to regain power. While the old conservative and Ultramontane parties still remained, a younger and more progressive wing was coming to the fore, grouped around the *Schweizerischen Studentenvereins* and the newspaper *Schwyer Zeitung*, led by Josef Gmür, brother of librarian Leonhard Gmür, who was Keller's contact in St Gall. During the conflicts preceding 1848, Karl Johann Greith had been one of the most outspoken champions of the Roman Catholic cause. He was an excellent orator and is sometimes referred to as the Bossuet of Switzerland.[37] Reeves showed his diplomatic wisdom by not taking sides with Keller and by keeping Greith involved in the transcription of the St Gall manuscript of the *Vita Columbae*. Karl Johann Greith sent him readings of St Gall MS 555 in April and in July 1853[38]. The two men corresponded in Latin and Greith's contribution is fully acknowledged in the *Life of St Columba*.[39] As it happened, instead of making an enemy, Reeves acquired a powerful friend since Greith was consecrated bishop of St Gall in 1862.

In his translation of Adomnán's *Vita Columbae* published in the Penguin Classics collection Richard Sharpe describes Reeves' book as 'a monumental edition'.[40] This is indeed true. It may be added that its significance goes beyond the excellent scholarly content and the classical, elegant production. The book epitomizes all that can be achieved with a spirit of resilience, optimism, goodwill and cooperation. The *Vita Columbae* was produced over a period when Reeves had many other pressing preoccupations. He was living in straightened financial circumstances as a curate in Ballymena, where he had to raise funds towards the building of a new church (consecrated 2 August 1852). He had to educate eight children and in 1855 suffered the death of his wife follow-

37 Victor Conzemius, 'Karl-Johann Greith', in *Dictionnaire d'histoire et de géographie ecclésiastique* 22 (1988, 104–14; Remo Wäspi, 'Carl Johann Greith (1863–1882). Der bedeutendste St Galler Bischof'. in Joachim Müller and Walther Baumgartner (eds) *Die Bischöfe des Bistums St Gallen* (Freiburg and Konstanz, 1996) pp. 61–84.

38 RIA, MS 12 W 29.

39 Reeves, *Life of St Columba*, p. xxvi.

40 R. Sharpe (ed.), *Adomnán, Life of St Columba* (London, 1995) p. 236.

ing the difficult birth of their ninth child. He did not know any German, yet he found in Ferdinand Keller a generous friend who made it possible for him to proceed with his edition. In exchange, he brought German scholarship to Ireland and organized the translation into English and the publication in the *Ulster Journal of Archaeology* of articles by Keller and Wilhelm Wattenbach.[41] At a time when confessional rivalry ran high, he managed in the process of publishing Adomnan's *Life of Columba*, not only to harness the cooperation of scholars of all persuasions, but also to make available to Catholics and Protestants the life of a saint who could be claimed as their own by both communities.

41 Wilhelm Wattenbach, 'Die Congregation der Schottenklöster in Deutschland' in *Zeitschrift für christliche Archaeologie und Kunst* (1856), 21–30 and 49–58, translated as 'The Irish monasteries in Germany' *Ulster Journal of Archaeology* [hereafter *UJA*] 1st ser. 7 (1859), 227–47; Ferdinand Keller, 'Bilder und Schriftzüge in den irischen Manuscripten der schweizerischen Bibliotheken', *Mittheilungen der antiquarischen Gesellschaft in Zürich* 7 (1851), 61–97, translated into English with an introduction and notes by Reeves as 'Early Irish Calligraphy', *UJA* 8 (1860), 210–30 and 291–308.

'An honour to the nation'

Publishing John O'Donovan's edition of the Annals of the Four Masters, *1848–56*

BERNADETTE CUNNINGHAM

A BANKRUPTCY SALE of the stock of Bryan Geraghty, bookseller and publisher, was held in Dublin at the Anglesea Street auction rooms of Charles Sharpe, on 29 February 1848.[1] Among the items to be disposed of were 327 copies of Owen Connellan's English translation of the Annals of the Four Masters, which Geraghty had published two years previously, although he had not yet paid the printer.[2] Connellan's translation of the annals for the years 1171–1616 was based on two transcripts of the text, one of which Connellan himself had made for Sir William Betham from the autograph manuscript acquired by the Royal Irish Academy in 1831.[3] The 736-page edition had been printed in Dublin by William Joseph Wilson, with illustrations designed, printed and engraved

1 *Catalogue of a very valuable and important collection of Irish manuscripts etc. ... to be sold by auction by Charles Sharpe, at his literary sale room, Anglesea Street, on Tuesday, 29th day of February, 1848, commencing at one o'clock* (Dublin, 1848).

2 Owen Connellan (ed.), *The annals of Ireland translated from the original Irish of the Four Masters* (Dublin, 1846); according to O'Donovan, Geraghty was responsible for the bankruptcy of the printer as well as himself (O'Donovan to John Windele, 25 June 1852, Royal Irish Academy, Dublin [hereafter RIA], MS 4 B 12/83 i).

3 Connellan (ed.), *The annals of Ireland*, preface. The Academy's autograph manuscript of the post-1171 annals is now bound in two volumes and numbered RIA, MSS 23 P 6–23 P 7.

by W.H. Holbrooke. Ornate capitals were used at the commencement of some pages of the text, apparently to enhance the appearance of each opening section of the text, which was originally planned as a periodical publication. The publisher explained that 'one of these ornamental letters has been given with each number of these annals, and will be so continued in every succeeding number, the publisher having for that purpose, at great cost, got all of them engraved by one of the ablest artists in Dublin, being anxious to preserve these curious and beautiful specimens of ancient Irish art, but few of which have ever been engraved in any printed works'.[4] The volume also contained a colour lithograph frontispiece, depicting four scholars at work in an implausibly elaborate gothic setting. Connellan's English translation of the Annals of the Four Masters was accompanied by extensive footnotes by Philip McDermott and Connellan himself. The notes rarely took the form of elucidations of the text, consisting mainly of general essays on aspects of Irish ecclesiastical and secular history adapted from the available secondary literature and only very tangentially related to the text of the annals.[5] The preface to the completed work admitted that the historical and topographical notes often had 'no immediate reference to the text', and explained that the serial nature of the original publication had created problems concerning the notes to the text. It was recorded, however, that the objective was 'to present to the Irish people as much as possible of their own history, at a moderate cost, and make it accessible to all'. Claiming to have more than 700 subscribers to the volume, Geraghty expressed his pleasure at having 'been enabled, through great exertions, to complete this national work, composed by the illustrious and learned O'Clerys, and to present to his countrymen, as a rare production of Irish

4 Connellan (ed.), *The annals of Ireland*, p. 145. There are no such ornamental capitals in the last 180 pages of the text, pp. 554–720, perhaps indicating that the idea of periodical publication had been abandoned.

5 See, for instance, the essay on south Connacht inserted at Connellan (ed.), *The annals of Ireland*, p. 121, n. 1. The note ends at p. 134 with the statement that 'The history and topography of the ancient kingdoms of Meath, Ulster, and Connaught, are finished in the present number, and in the succeeding numbers will be given the history and topography of Leinster and Munster', a further indication of the intention to publish the work in sections over a period of time.

genius and patriotism, and lost to the literary world for more than two hundred years; and he confidently hopes, that his future undertakings in Irish literature will receive the support of all patriotic and enlightened Irishmen'.[6]

Not all patriotic Irishmen were enthused, however, by Geraghty's publishing endeavours, even despite their interest in the text of the annals. For instance, William Hackett had sought to borrow the Cork antiquarian John Windele's copy of the serial edition in February 1845. Having enquired whether Windele was 'taking the cheap Four Masters', Hackett went on to say it was 'a pity to encourage such an invidious project but, as I would not consider my borrowing it from you would be any benefit to the Publishers, I should not scruple you sending it (or my receiving it rather) by some convenient opportunity.'[7] By 1845, Hackett, Windele, and other potential purchasers were well aware that another edition of the Annals of the Four Masters, being prepared by John O'Donovan, was well advanced,[8] and they regarded the appearance of a rival cheap edition as being somewhat inappropriate. At any rate, despite his claim to have 700 subscribers, the publication appears to have been the immediate cause of Geraghty's bankruptcy in 1848.[9]

Given the fate of Connellan's edition, it must have been with considerable trepidation that Messrs Hodges and Smith, of 104 Grafton Street, received the sheets of their three-volume, dual-language edition of the very same annals from M.H. Gill of Dublin University Press in the same spring of 1848. Retailing at 8 guineas (6 guineas to subscribers)[10] as compared with the 30 shillings[11] charged by Geraghty for Connellan's one-volume English translation, the Hodges and Smith venture must

6 Connellan (ed.), *The annals of Ireland*, preface.
7 William Hackett to John Windele, 6 Feb. 1845 (RIA, MS 4 B 5/88).
8 Hodges and Smith's formal prospectus for the Annals had been circulated to potential subscribers in January 1844.
9 John O'Donovan to John Windele, 25 June 1852 (RIA, MS 4 B 12/83 i).
10 *Preparing for publication, in two volumes, 4to., The Annals of the Four Masters, from AD 1172, to the conclusion, in 1616, consisting of the Irish text, from the original MS, and an English translation with copious explanatory notes, by John O'Donovan, Esq. Prospectus of the work* (Dublin, 1844), p. 9 (RIA, 12 I 15, pp. 319–27).
11 RIA, Council minutes, VII, p. 210.

have been a high-risk one. However, the publishers had already invested heavily in the project well before Connellan's 'cheap' edition appeared, and the edition they planned was considerably more ambitious.

EARLY PUBLICATION PLANS

The impetus to publish the AD 1172–1616 section of the Annals of the Four Masters dates from the time the only complete autograph manuscript of that part of the annals was acquired by the Royal Irish Academy in 1831.[12] Another autograph copy was already in Trinity College, Dublin, but it was incomplete, the AD 1172–1335 section and the closing entry for 1616 having been lost.[13] An eighteenth-century transcript was also already in the Royal Irish Academy,[14] but the excitement of acquiring the original manuscript in the hands of the Four Masters sparked immediate interest in having the work published. The autograph manuscript had been bought by George Petrie for the Academy at the sale of Austin Cooper's books, on 21 February 1831. Petrie paid £53 for it, and gave it to the Academy for the same price.[15] By 7 March 1831 a special committee[16] had reported to the Council of the Academy with its recommendations for translating and publishing the text of the Annals. Among the issues addressed at this planning stage was whether the Irish text should be translated into English or Latin. An edition of the earlier portion of the annals to AD 1171, published privately by the duke of Buckingham in 1826, had contained a Latin translation prepared by the Revd Charles O'Conor.[17] The Academy committee foresaw a difficulty

12 RIA, MSS 23 P 6–23 P 7 (cat. nos. 687–688).

13 Trinity College, Dublin [hereafter TCD], MS 1301 (formerly H.2.11).

14 RIA, MS 23 F 4 (cat. no. 990) contained a transcript by Muiris Ó Gormáin of the annals for the years AD 1171–1397.

15 RIA, Council minutes, IV, p. 295; William O'Sullivan, 'The Slane manuscript of the Annals of the Four Masters', *Ríocht na Mídhe*, 10 (1999), 78, 84, citing a priced copy of the sale catalogue (TCD, Gall. A. 6. 54).

16 The committee comprised John D'Alton, Casar Otway, W.H. Drummond and George Petrie.

17 Revd Charles O'Conor (ed.), *Quatuor Magistrorum annales Hibernici usque ad annum MCLXXII ex ipso O'Clerii autographo in Bibliotheca Stowense servato* (Rerum

in finding any scholar with the necessary expertise in reading the Irish manuscript who would also be competent to produce a Latin translation. They wisely rejected the idea of having the text translated from Irish to English and then from English into Latin, 'to prevent the errors which would, it is supposed, be incident to versions by two distinct individuals into two different languages'.[18] With an eye to the market it was recommended that the Annals be published with an English translation, in the belief that the proposed work 'would so obtain a more extensive sale, and thereby more widely diffuse a taste for Irish literature, and enable the Academy when the same was completed, to proceed in the publishing of others of great national importance.'[19] In May 1831 a new Academy committee was appointed, this time to 'make prompt enquiries as to the facility of obtaining a competent individual to translate the said annals into scholarlike English' and also to consider the likely cost and time frame of the project.[20]

By October 1837 the Academy had concluded that the total cost would be in the region of £3,000. This sum included not just printing and production costs but also payments to scholars to prepare a full English translation and to have the Irish text transcribed in a form that could be sent to the printer.[21] The Academy sought government funding for the project, explaining that of all the categories of Irish manuscripts that were worthy of publication, priority was being given to the Annals because they presented the least technical difficulties and, when published, would provide useful historical and lexicographical materials to assist in editing other sources, legal, literary, ecclesiastical and genealogical.[22] The Academy's request for funding for the project was rejected by the chancellor of the exchequer. It was becoming clear that if publication was to proceed, the work would have to be financed by a commercial publisher.

Hibernicarum Scriptores Veteres III, Buckingham, 1826). Later commentators praised the generosity of the duke of Buckingham in financing this work, but also pointed to the 'glaring errrors' of that edition (*Irish Quarterly Review*, 1 (1851), 694–5).

18 RIA, Council minutes, IV, pp. 295–6.

19 Ibid., pp. 295.

20 Ibid., pp. 304–5.

21 RIA, Council minutes, V, pp. 54–60.

22 Ibid., pp. 54–7.

GEORGE SMITH

Mr George Smith had become active as a bookseller from 1829, operating as a partner in a commercial business under the name Hodges and Smith, successor to Hodges and McArthur, of 21 College Green, Dublin. Hodges and McArthur had been formally recognized as 'booksellers to the Academy' in 1818, and by the 1840s Hodges and Smith had obtained permission from Trinity College, Dublin, to describe themselves as 'booksellers to the University'.[23] They also moved a short distance to new premises at 104 Grafton Street, close to the Royal Irish Academy then located at 114 Grafton Street opposite the provost's house of Trinity College. It was a substantial shop, rated at £180 as compared with £120 for the building occupied by the Academy.[24] Among the other businesses on Grafton Street in 1850 describing themselves as publishers as well as booksellers were Thomas Cranfield (no. 115), John H. Powell (no. 88), William Carson (no. 51), and John Robertson (no. 3), while Samuel James Pigott (no. 112) was a specialist music publisher. Among the more specialist booksellers and stationers on Grafton Street, John Fannin (no. 41) specialized in medical books, James Cornish (no. 18) in theology while Charles Hedgelong (no. 20) dealt in second-hand books, and George Bellew (no 79) was a bookbinder as well as bookseller. Other general booksellers with premises on the same street were Edward Ponsonby (no. 116), Joshua Porter (no. 72), W.B. Kelly (no. 8) and Madden and Hare (no. 7), the last two located directly opposite Hodges and Smith's new premises, at the Trinity College end of the street.[25]

George Smith, though never a member of the Royal Irish Academy himself,[26] was closely involved with the antiquarian work being spearheaded by Academy members such as George Petrie and James Hardiman. His efforts – and his money – did not always draw the appre-

23 *Works on Ireland published by Hodges and Smith, booksellers to the University, 104, Grafton Street, Dublin* (London & Dublin, 1848).

24 *Thom's Irish almanac and official directory ... 1850* (Dublin, 1850), pp. 736–8.

25 *Thom's directory ... 1850*, pp. 736–8.

26 He was not mentioned in the history of the Academy published to mark its bicentenary, T. Ó Raifeartaigh (ed.), *The Royal Irish Academy: a bicentennial history, 1785–1985* (Dublin, 1985).

ciation he expected from the Academy and he was quite indignant when in 1848 he was refused permission to borrow artefacts from the Academy museum for the purpose of having illustrations made for the catalogue of the museum's holdings then in preparation.[27] In his letter of complaint he pointed out that he had financed the acquisition of some gold torques, to save them from being melted down, and had himself travelled more than 200 miles to obtain the cross of Cong for the Academy, and had advanced the money to pay for it.[28] The Academy eventually compromised, allowing George Petrie to borrow and be responsible for the safe return of the artefacts requested by Smith.[29]

In his early years in the book trade George Smith assembled a collection of 227 volumes of Irish manuscripts, and these he had offered for sale to the Academy in 1843 for the considerable sum of 1,250 guineas. Smith had employed Eugene O'Curry to prepare a detailed catalogue of collection, which mainly comprised eighteenth- and nineteenth-century manuscripts, but also included such treasures as the twelfth-century *Leabhar na hUidhre*.[30] It was an attractive offer, despite the high price, and the Academy eventually succeeded in raising the funds necessary to purchase the Hodges and Smith manuscripts through private subscriptions from members matched pound for pound by a government grant to supplement the Academy's own meagre funds.[31]

In addition to their role as sellers of books and manuscripts, Hodges and Smith's involvement with the Academy as publishers continued well into the 1860s, and the company was responsible, for example, for the publication of William Wilde's impressive illustrated catalogues of the antiquities in the Academy museum.[32] Other prestige items in which they were involved included George Petrie's major study of Irish round

27 RIA, Council minutes, VII, pp. 277–8.

28 Ibid.

29 RIA, Council minutes, VII, p. 278.

30 O'Curry's three-volume manuscript catalogue of the Hodges and Smith collection was acquired by the Royal Irish Academy along with the collection.

31 R.B. McDowell, 'The main narrative', in T. Ó Raifeartaigh (ed.), *The Royal Irish Academy: a bicentennial history*, p. 43; RIA, Council minutes, VI, pp. 198, 200, 204, 208, 212–16, 217, 233.

32 W.R. Wilde, *A descriptive catalogue of the antiquities in the museum of the Royal Irish Academy: articles of stone, earthen, vegetable, and animal materials; and of copper and*

towers and ecclesiastical architecture,[33] a co-publication venture with the Royal Irish Academy, the first edition of the work having been published in 1835 as volume 20 of the *Transactions of the Royal Irish Academy*.[34] An arrangement was made in 1845 that Hodges and Smith would print 450 copies of Petrie's *Round towers* bearing the Academy imprint and a further 300 with their own imprint, though when 500 were accidentally printed with the Academy imprint, and just 250 with the Hodges and Smith imprint, the Academy agreed to take the extra fifty copies on the same terms as the other 450.[35] Hodges and Smith also acted as publishers for the Ordnance Survey, Dublin University, the Queen's Colleges at Cork and Galway, the Royal Dublin Society, and the Dublin Statistical Society. While the bulk of their publishing through mid-century was in the academic, legal and ecclesiastical fields, Hodges and Smith also produced the occasional work of fiction.[36] Irish-language works published by Hodges and Smith included a revised edition of John O'Brien's Irish-English dictionary, originally published in 1768,[37] and John O'Donovan's Irish grammar, designed for school use.[38]

The publication of the Annals of the Four Masters was probably George Smith's most ambitious business venture: it was one of his earliest major publishing commitments and was what made his reputation. Thus, in a cameo appearance in Lady Ferguson's late nineteenth-centu-

bronze (Dublin and London 1857; reprint Dublin, 1863); W.R. Wilde, *A descriptive catalogue of antiquities in gold in the museum of the Royal Irish Academy* (Dublin and London, 1862). (Printed by M.H. Gill at the University Press, these catalogues were co-published by Williams and Norgate of London, and while the title page named only the commercial publishers, the cover, of some copies at least, included the Royal Irish Academy in the imprint.)

33 George Petrie, *The ecclesiastical architecture of Ireland anterior to the Anglo-Norman invasion, with an essay on the round towers of Ireland* (Dublin, 1845).

34 Petrie had been awarded a gold medal prize by the Royal Irish Academy for the work in 1833.

35 RIA, Council minutes, VII, pp. 10–11.

36 [Charles Bernard Gibson], *The last earl of Desmond: a historical romance of 1599–1603 in two volumes*, 2 vols (Dublin, 1854).

37 John O'Brien, *Focaloir Gaoidhileg-Sax-Bhearla, or an Irish-English dictionary* (2nd ed., Dublin, 1832).

38 John O'Donovan, *A grammar of the Irish language published for the use of the senior classes in the College of St Columba* (Dublin, 1845).

ry biography of Samuel Ferguson, he was introduced as 'Mr George Smith, the eminent publisher of the "Annals of the Four Masters"'.[39] Smith had a long-standing association with the scholarly circle that included George Petrie, Thomas Larcom, J.H. Todd, William Stokes, John O'Donovan and Eugene O'Curry. Among the key activities that drew these men together were the topographical work of the Ordnance Survey and the initiative of the Irish Archaeological Society in publishing editions and translations of Irish language texts of antiquarian interest. Recalling this circle of scholars in later life, Samuel Ferguson noted 'All the men named, and many more who might be named, in the high walks of thought and practice, were the friends and intimates of George Smith ... He associated with them not merely as a publisher, but as a sympathiser in all their aims and ambitions'.[40] Smith's decision to become involved with publishing the Annals emerged from this context, influenced in no small way by the enthusiasm for the project displayed by George Petrie. Ferguson remembered the 'gentle exaltation' shown by Petrie at a dinner party in Smith's house when the publication of the annals was decided on.[41]

GEORGE PETRIE AND THE TYPE DESIGN FOR THE ANNALS

From the time he purchased the autograph manuscript of the Annals of the Four Masters for the Academy in 1831, George Petrie was anxious that it should be published. Petrie had stressed 'the necessity of giving durability, while yet in our power, to the surviving historical remains of our country, and thereby placing them beyond the reach of a fate otherwise inevitable. To me it appears a sacred duty on all cultivated minds to do so.'[42] Petrie had a clear idea of the kind of edition required, and was particularly keen that the Irish text should be printed using an appro-

39 Lady Ferguson, *Sir William Ferguson in the Ireland of his day*, 2 vols (Edinburgh, 1896), i, p. 69.
40 Ferguson, *Sir William Ferguson*, i, p. 70.
41 Ibid., i, p. 69.
42 George Petrie, 'Remarks on the history and authenticity of the Annals of the Four Masters', *Transactions of the Royal Irish Academy*, 16 (1831), 381–93, at p. 387.

priate Gaelic font. The inadequacies of the italic type used in Charles O'Conor's edition of the earlier part of the annals had been emphasized by the Academy when seeking government funding for their proposed edition:

> The Irish text has been printed in Italic characters, and the numerous contractions of the original manuscript have been retained, without any attempt to represent, by proper type, the significant ligatures on which their meaning entirely depends. Hence, Dr O'Conor's text is of little use to the student of Irish, and is, in many places, unintelligible, except to the most practised scholars. We would therefore propose to publish the Irish text in the antient character, and without contractions, and to accompany it with an English version.[43]

By 1835 George Petrie had designed a new Gaelic type for use by the Dublin University Press. Petrie's 'A' type, funded by Hodges and Smith, was modelled on the lettering in the Book of Kells.[44] It was first used in the *Ordnance Survey of County Londonderry: I, the parish of Templemore*, printed by the University Press and published for the government by Hodges and Smith in 1837, but Dermot McGuinne has argued that the type was originally planned in 1832–3 when the publication of the Annals of the Four Masters was first proposed.[45] The type was subsequently used in articles by Petrie containing Irish text published in the *Transactions of the Royal Irish Academy*, and was also selected for use in the earliest publications of the Irish Archaeological Society.[46] Petrie's 'A' type, for-

43 RIA, Council minutes, V, p. 58.

44 Vincent Kinane, *A history of the Dublin University Press, 1734–1976* (Dublin, 1994), pp. 130–1.

45 Dermot McGuinne, *Irish type design: a history of printing types in the Irish character* (Dublin, 1992), pp. 102–3. Both Petrie and O'Donovan were closely involved in the preparation of the Londonderry volume.

46 McGuinne, *Irish type design*, pp. 105–6; *Tracts relating to Ireland*, I (Dublin, I.A.S., 1841). M.H. Gill of Dublin University Press financed a second type designed by Petrie in 1850. The use of this 'B' type was restricted to works printed at the Dublin University Press, so that even Petrie himself was only permitted to use it if his work

mally named 'Irish Hibernian' type, was copyrighted by John Hodges and George Smith of 21 College Green in April 1844,[47] shortly after the publisher's prospectus for the Annals was first circulated.[48]

The design and physical appearance of the Irish text of the Annals was clearly regarded as important. While the use of a Gaelic rather than italic script was readily justifiable on scholarly grounds, as necessary for the accurate representation of the written text, its significance transcended those technicalities. Just as the Gaelic origins of Irish place-names were being brought to light by O'Donovan and other researchers in the course of the Ordnance Survey cartographic project, so the Gaelic texts that were central to Irish antiquarian research were being brought into the public sphere through the medium of print.[49] Their Gaelic character was what defined the authenticity of these texts, and it was deemed necessary to preserve the essence of that character in the print medium. It was all part of the process of Gaelicizing the memory of the past.[50] Following the lead given by Charlotte Brooke's *Reliques of Irish poetry* (1789), parallel Irish and English texts on facing pages was adopted as a means of visibly demonstrating the authenticity of texts.[51] The same policy was adopted in the publications of the Irish Archaeological Society, founded in 1840. Thus, whereas Connellan's 'cheap' serial edition of the Annals of the Four Masters had settled for the English translation on its own, such an approach would not have been considered adequate by Petrie and his associates.

was printed by the University Press (Kinane, *Dublin University Press*, p. 131; McGuinne, *Irish type design*, pp. 103, 119).

47 McGuinne, *Irish type design*, pp. 102–4.

48 The circular letter accompanying the first prospectus was dated 31 Jan. 1844 (RIA, 12 I 15, p. 311); a reminder was dated 2 Mar. 1844 (RIA, 23 I 15, p. 315). P. Walsh, *The Four Masters and their work* (Dublin, 1944), pp. 35–8.

49 Joep Leerssen, *Hidden Ireland, public sphere* (Galway, 2002). pp. 24–7. Leerssen points out that the research into Irish place-names overseen by the Ordnance Survey was 'disastrously misrepresented' in Brian Friel's play, *Translations* (Leerssen, *Hidden Ireland*, p. 45, note 11). For a systematic defence of O'Donovan's onomastic scholarship see J.H. Andrews, '"More suitable to the English tongue": the cartography of Celtic place-names', *Ulster Local Studies*, 14:2 (1992), 7–22, at pp. 18–19.

50 Or, as Leerssen describes it, the 'deanglicization of Ireland's public sphere' (Leerssen, *Hidden Ireland*, p. 39).

51 Ibid., p. 25.

As Joep Leerssen has noted, the medium of print was used to move the cultural heritage of Gaelic Ireland from its restricted manuscript context and relocate it in the public sphere.[52] In a scholarly atmosphere where particular importance was attached to collecting antiquarian artefacts and manuscripts, publication of texts was a logical further step in the process of bringing these records of Irish heritage into public ownership, a process in which the Royal Irish Academy had been engaged since its foundation in 1785, and which accelerated greatly in the 1840s when the Academy purchased some major manuscript collections.[53] The belief in the national importance of publishing such manuscripts let to the formation of the Irish Archaeological Society, founded on St Patrick's Day 1840, under the leadership of J.H. Todd specifically to arrange for the publication of editions and translations of manuscripts of particular Irish historical significance.[54]

In Hodges and Smith's 1844 prospectus for their proposed edition of the Annals of the Four Masters, there is passing reference to a suggestion that the task of publication should be left in the hands of the Irish Archaeological Society. Hodges and Smith pointed out that such an aspiration was unrealistic on financial grounds. 'Now that the Work is ready for the Press, were we to transfer it to the Society, the state of its funds and numerous engagements would necessarily delay the Publication for many years.'[55] Despite the financial implications, the Academy also seems to have continued to harbour an ambition to become the publisher of the annals, applying for government funding

52 Ibid., p. 39; see also Damien Murray, *Romanticism, nationalism and Irish antiquarian societies, 1840–80* (Maynooth, 2000), pp. 96–123.

53 Elizabeth FitzPatrick, *The catalogue of Irish manuscripts in the Royal Irish Academy: a brief introduction* (Dublin, 1988), pp. 3–5. The collection of museum artefacts by the Royal Irish Academy, a project also spearheaded by Petrie, and sometimes financed by George Smith, was another expression of the same ideology. Joep Leerssen, *Remembrance and imagination: patterns in the historical and literary representation of Ireland in the nineteenth century* (Cork, 1996), pp. 106–7; G.F. Mitchell, 'Antiquities', in Ó Raifeartaigh (ed.), *The Royal Irish Academy: a bicentennial history*, pp. 93–163.

54 For an analysis of the membership of the Irish Archaeological Society see Murray, *Romanticism, nationalism and Irish antiquarian societies*, pp. 62–7.

55 Hodges and Smith circular letter, 31 Jan. 1844 (RIA, 12 I 15, p. 311).

for that very project as late as 1837.[56] Since Petrie knew of the Hodges and Smith plan from its inception in the early 1830s, it is unclear why the Academy was still attempting to pursue this option. It seems that some Academy members may have deemed publication by the Academy as preferable to commercial publication in the case of a text of such national significance.

O'DONOVAN'S WORK

When George Smith made an arrangement with John O'Donovan to translate and edit the annals,[57] it appears that he commenced work straight away in January 1832.[58] By May 1833 he had completed the 'first volume'[59] and noted that he planned to 'rest for some time' before resuming the task of translating the second volume.[60] From the autumn of 1833 to the end of 1842 John O'Donovan was employed on a full-time basis in the topographical department of the Ordnance Survey; the department was headed by none other than George Petrie. He had first begun work for the Ordnance Survey, at a lower grade, in September 1830, resigned in January 1833, and was reappointed at a higher salary later that year. Petrie's high praise of O'Donovan in the *Dublin Penny Journal* of 8 June 1833 as 'the most able and judicious Irish scholar and topographer Ireland has produced for the last century'[61] may have been designed to support O'Donovan's quest for better employment terms. During the next nine years O'Donovan spent much of his time travelling the country conducting topographical and onomastic research. While this was done primarily to assist the cartographic work of the Ordnance Survey, it is evident that O'Donovan's extensive topographi-

56 RIA, Council minutes, V, pp. 64–70.
57 Hodges and Smith circular letter, 31 Jan. 1844 (RIA, 12 I 15, p. 311).
58 Extract from *Dublin Review*, 24 (1848), cited in 'Opinions of the Press' (RIA, 12 I 15, p. 588).
59 RIA, MS 23 P 6.
60 Art Ó Maolfaobhail, 'Éadbhard Ó Raghallaigh, Seán Ó Donnabháin, agus an tSuirbhéireacht Ordanáis', *Proceedings of the Royal Irish Academy*, 91 C (1991), 73–103, at p. 98.
61 *Dublin Penny Journal*, 8 June 1833, p. 396.

1 John O'Donovan

cal notes that accompany his edition of the Annals were assembled during the years that he was actively engaged in research throughout the country. The publishers made a virtue of the fact that O'Donovan had visited virtually every townland in the country in the course of his research.[62] The disbandment of the topographical department at the end of 1842 meant a change in his employment status although he was retained as an adviser to the Ordnance Survey down to the time of his death in 1861.[63] The change gave O'Donovan the opportunity to spend more time in the library of the Royal Irish Academy, and to devote his energies to editing Irish manuscripts. The various publications of the Irish Archaeological Society through the 1840s reveal, however, that this time was devoted to editing texts other than the Annals of the Four Masters.[64] It seems probable, therefore, that the work of translating and

62 Circular letter from Hodges and Smith, 31 Jan. 1844 (RIA, 12 I 15, p. 311).

63 Nollaig Ó Muraíle, 'John O'Donovan', *Oxford dictionary of national biography*, 60 vols (Oxford, 2004), xli, pp. 527–9; see also, Nollaig Ó Muraíle, 'Seán Ó Donnabháin: an cúigiú máistir', in *Scoláirí Gaeilge: Léachtaí Cholm Cille XXVII*, ed. Ruairí Ó hUiginn (Maynooth, 1997), pp. 11–82.

64 Works translated and edited by John O'Donovan and published by the Irish

annotating the Annals was essentially complete by January 1844, when the publishers first formally announced their plans for publication. Indeed, the work of translation had been completed more than a decade earlier.[65] Work remained to be done in planning the detail of the format in which the text would be published, and for this purpose an advisory committee comprising J.H. Todd, George Petrie, Aquilla Smith and Joseph Huband Smith was formed in February 1845.[66]

By this stage O'Donovan himself had considerable editorial experience, having been employed for some years to edit and translate texts for the Irish Archaeological Society. His prodigious output for that body, his principal employer in the 1840s, was impressive, though he sometimes found the work wearisome. He complained to William Mooney in 1845 that 'I have so many "irons in the fire" that I am just sick to death's door of books and writing. The Blacks may complain of slavery, but here I sit a white *subject*, a greater slave than any of them'.[67] Arguably O'Donovan's greatest achievement, his Annals of the Four Masters was just one more 'iron in the fire', undertaken not merely for the love of scholarship, but also as a means of supporting his wife and family. Just as the Irish Archaeological Society paid him for the work he undertook under Todd's direction, so in the case of the Annals of the Four Masters he was paid by George Smith.[68] In the same way, Eugene O'Curry was

Archaeological Society during the 1840s include *The banquet of Dún na n-Gédh and the Battle of Magh Rath, an ancient historical tale* (1842), *The tribes and customs of Hy-Many, commonly called O'Kelly's country* (1843), *The genealogies, tribes, and customs of Hy-Fiachrach, commonly called O'Dowda's country* (1844). He also contributed editions of shorter texts to two miscellaneous volumes, *Tracts relating to Ireland*, vol. 1 (1841), and *The miscellany of the Irish Archaeological Society*, vol. 1 (1846). His edition of *Leabhar na gCeart, or the book of rights* (1847) was published by the Celtic Society, and he also edited the *Miscellany of the Irish Celtic Society* (1849).

65 *Gentleman's Magazine*, n.s. xx (1848), p. 385.

66 John O'Donovan (ed.), *The Annals of the Kindgom of Ireland by the Four Masters* (2nd ed., 7 vols, Dublin, 1856), i, pp. xxxviii-xxxix.

67 John O'Donovan to William Mooney, 21 Aug. 1845, cited from Ordnance Survey files recently transferred to the National Archives of Ireland, in Ó Muraíle, 'Séan Ó Donnabháin: an cúigiú máistir', p. 70.

68 O'Donovan recorded payments totalling £150 received from George Smith between January 1848 and May 1849, usually in monthly payments of £10. (Colm O'Lochlainn, 'John O'Donovan and the annals', *Irish Book Lover*, 27 (March 1940), 179). This was probably payment for his work on volumes one and two of the

paid by the publishers to produce a full transcript of the autograph manuscript of the Annals of the Four Masters.[69]

In February 1844, Hodges and Smith was granted permission by the Royal Irish Academy to have a transcript made of the autograph manuscript of the Annals of the Four Masters.[70] The work of transcription, presenting the text in a form suitable for use by the printers at the University Press, was to be done by Eugene O'Curry in the Academy. When the completed work was eventually published, O'Donovan acknowledged the contribution of O'Curry,

> by whom the autograph of this work was copied for the Press, and who has supplied very many examples from ancient glossaries to elucidate the meanings of difficult words, and various manuscript authorities, unexplored by any but himself, to illustrate the ancient topography.[71]

Placed half way through the fifty-third of fifty-four pages of preliminary matter, it seems small acknowledgment for the considerable amount of work undertaken by O'Curry in connection with the edition.

MARKETING THE FIRST EDITION

The prospectus for their edition of the Annals of the Four Masters issued by Hodges and Smith at the end of January 1844 was intended to attract a minimum of 200 subscribers.[72] Four weeks later, the names of eighty-one subscribers were circulated and by the middle of March a total of 145 firm orders had been received.[74] The circular letter, which accom-

extended edition published in 1851. The total sum paid to O'Donovan for his work on the Annals is not known.

69 RIA, Council minutes, VI, p. 218. 70 Ibid., p. 218.

71 O'Donovan (ed.), *The annals of the kingdom of Ireland* (2nd ed. 1856), i, p. liii.

72 Printed circular letter from Hodges and Smith, 31 Jan. 1844 (RIA, 12 I 15, p. 311).

73 The preliminary subscription list of eighty-one persons, twenty-one of whom were members of the Academy, was used for publicity purposes (RIA, 12 I 15, pp. 313–14).

74 Hodges and Smith to John Windele, 19 Mar. 1844 (RIA, MS 4 B 5/16).

panied the prospectus, made reference to the long-standing nature of the arrangement between the publishers and the editor.

> From the time that the Royal Irish Academy first became possessed of the Autograph Copy of the Annals we felt an anxious desire to publish them, and were only delayed by the impossibility of finding an Editor competent to undertake the Translation and give explanatory Notes, particularly these identifying the ancient and modern topography of Ireland; feeling that this difficulty was likely to be removed when Mr. O'Donovan was appointed by the Head of the Ordnance Survey in Ireland to collect topographical information, and correct the orthography of the Townland Names, we entered into an arrangement with him to undertake the Editorship. As almost every Townland throughout Ireland required to be visited by Mr. O'Donovan, we knew that several years would elapse before the Publication could be properly ready for the Press, but we felt that in bringing before the Public so great a national Work, that time and expense should be secondary considerations. Mr. O'Donovan has lately finished his investigations, and we feel confident that the extraordinary mass of information given in the Notes will be considered of vast importance in a topographical and literary point of view by all persons interested in the history of Ireland.[75]

While O'Donovan may not originally have envisaged the publication of the results of his topographical researches in the form of notes to the Annals, the termination of the Ordnance Survey's topographical project meant that there was no other realistic possibility of his publishing the bulk of the work he had been engaged on in the decade prior to 1842.

At the time the forthcoming publication of the Annals was first announced by Hodges and Smith in 1844 just two printed volumes were envisaged, with a print-run of 500 copies. By the time the work was

75 Printed circular letter from Hodges and Smith 31 Jan. 1844 (RIA, 12 I 15, p. 311).

printed in 1848, John O'Donovan's very extensive notes to the text meant that the work had grown to some 2,498 pages which necessitated it being issued in three volumes. The three-volume set covered the years 1172–1616, and were introduced to prospective customers as a continuation of the text of the early part of the Annals of the Four Masters, A.M. 2242–AD 1171, that had been privately published by the duke of Buckingham in 1826. Hodges and Smith's prospectus explained that 'this hitherto unpublished portion, extending from AD 1172 to AD 1616, and of which the original authorities are now, in great part, lost, comprises more than three-fourths of the entire compilation, so that the proposed publication may be regarded as virtually giving these Annals to the world for the first time'.[76] No subscription list was published in the edition of the Annals of the Four Masters issued by Hodges and Smith in 1848, but sufficient orders were received for the work to be printed.[77] The name of each individual subscriber was printed on the verso of the title page. Among those who had shown a marked reluctance to subscribe was the Cork antiquarian, John Windele, who informed Hodges and Smith in 1844 that while he intended to purchase the volume on publication, he would not take out a subscription in advance because of earlier difficulties in respect of subscriptions to the *Topographical dictionary of Ireland* published in London by Samuel Lewis.[78] Hodges and Smith took the trouble to notify Windele that by declining the subscription offer he would have to pay the higher price of 8 guineas, when the work was published, but Windele did not relent.[79] Indeed, his unwillingness to subscribe was still a talking point some eight years later.[80]

76 *Annals of the Four Masters: prospectus of the work* [1844], p. 2.

77 Subscribers to the original 1848 edition had the option of two different qualities of paper, and it may be that the higher quality product initially sold better. On the choice of format available to original subscribers in Oct. 1847 see RIA, Council minutes, VII, p. 242.

78 Samuel Lewis, *A topographical dictionary of Ireland* 3 vols (London, 1837). The first edition of this work, which was published by Samuel Lewis, included a very extensive subscription list.

79 Hodges and Smith to John Windele, 19 Mar. 1844 (RIA, MS 4 B 5/16).

80 John O'Donovan to John Windele, 25 June 1852 (RIA, MS 4 B 12/83 i).

Annala Rioghachta Eireann.

ANNALS

OF

THE KINGDOM OF IRELAND,

BY THE FOUR MASTERS,

FROM

THE EARLIEST PERIOD TO THE YEAR 1616.

EDITED FROM MSS. IN THE LIBRARY OF THE ROYAL IRISH ACADEMY AND OF TRINITY COLLEGE, DUBLIN, WITH A TRANSLATION, AND COPIOUS NOTES,

BY JOHN O'DONOVAN, LL. D., M. R. I. A.,

BARRISTER AT LAW.

" Olim Regibus parebant, nunc per Principes factionibus et studiis trahuntur: nec aliud adversus validissimas gentes pro nobis utilius, quam quod in commune non consulunt. Rarus duabus tribusve civitatibus ad propulsandum commune periculum conventus: ita dum singuli pugnant universi vincuntur."—TACITUS, AGRICOLA, c. 12.

VOL. I.

DUBLIN:
HODGES AND SMITH, GRAFTON-STREET,
BOOKSELLERS TO THE UNIVERSITY.
1851.

2 Title page of the first volume of the seven-volume set of the *Annals of the Kingdom of Ireland* issued in 1851

THE COMPLETE EDITION OF 1851

Although received with critical acclaim, O'Donovan's three-volume 1848 edition of the Annals of the Four Masters was not a commercial success. Being published during the Famine years, and at a time of political unrest, all concerned were well aware that it was an inauspicious time for such an enterprise. Reflecting on the project in 1852, O'Donovan observed:

> my own opinion is that Ireland was not in a condition at the time <i.e. 1845 to 1850, awful period!!> to bring out so heavy a work, and that therefore loss was inevitable, but it cannot be helped now.[81]

Despite the lack of commercial success of the first edition, the publishers decided to expand the project and commissioned O'Donovan to produce an edition of the early part of the annals, to AD 1171, in the same format, together with an index to the complete work. The publishers asserted that 'we never undertook the publication with a view to profit; and even now all we can expect from the publication of the completion of the work will be a diminution of our very heavy loss.'[82] The extended edition, now comprising six volumes of text and a seventh volume containing an index, was published in 1851. The binding was 'designed from the elaborate case of the shrine of St Maidoc, or Aidan, the first bishop of Ferns'.[83] The first two volumes of the 1851 edition contained a previously unpublished edition and translation, with O'Donovan's topographical notes, of the annals from AM 2242 to AD 1171. To ensure uniformity of size of the volumes, the original three volumes of the 1848 edition, from AD 1172 to 1616, were now issued as four volumes. The unbound sheets left unsold from the 1848 edition were used to make up the new volumes three to six, with new title pages being printed as neces-

81 John O'Donovan to John Windele, 25 June 1852 (RIA, MS 4 B 12/83 i); the phrase in angle brackets was interlined.

82 *The completed Annals of the Four Masters* [*private circular*], p. 2 (RIA, 12 I 15, p. 364).

83 Described by the reviewer writing for the *Irish Quarterly Review*, 1 (1851), 697n.

sitated by the revised format. Subscribers who had already purchased the 1848 edition were offered a special deal on volumes one and two and the index volume and also received the revised title pages together with instructions to the binder on the make-up of the seven-volume edition.[84] Thus, as Vincent Kinane has shown, neither the 1851 edition nor indeed the 1856 'second' edition were new printings, but rather schemes that made use of the sheets remaining unsold from the original 1848 printing.[85] Although the errata noted in the 1848 printing remained uncorrected, the outcome in 1851 was an enhanced seven-volume product that was a complete edition of the text as then known and which also contained invaluable indexes of names and places.[86] When the irrepressible John Windele complained that the index volume did not extend to an index of themes, John O'Donovan replied unapologetically that 'I found the index to the Annals exceedingly laborious. It was in fact too heavy for any one person. An index to the miscellaneous subjects would have killed me, and I am not sorry for not having attempted it.'[87] In the very next sentence, undaunted, O'Donovan went on to mention that 'the next work we have in view is the law of the Brehons, which will be the heaviest yet attempted. If we accomplish it I trust it will redound to the glory of our own *poor* times', and also casually mentioned that he was assisting John O'Daly in his work on another early text 'from a kind of innate love of the subject'.[88]

Sales of the seven-volume edition must have been rather less brisk than anticipated and John O'Donovan reported to John Windele in June 1852 that he was unwilling to negotiate with Hodges and Smith to get some loose sheets so as to make up a set of the Annals, adding 'they have lost so much by this publication that they can hardly bear the sight of me in their shop.'[89] The price for the complete edition was a very sub-

84 See, for example, J.H. Todd's set of the 1851 edition in the Bradshaw Collection at Cambridge University Library, which includes the instructions to the binder.

85 Kinane, *Dublin University Press*, pp. 154–6.

86 Hodges and Smith announced that there would be 250 copies of the 1851 edition (RIA, 12 I 15, pp. 300, 306).

87 John O'Donovan to John Windele, 17 Dec. 1851 (RIA, MS 4 B 11/108).

88 Ibid.

89 John O'Donovan to John Windele, 25 June 1852 (RIA, MS 4 B 12/83 i). O'Donovan

stantial 14 guineas, a price which the publishers well knew 'chiefly confined the sale to public institutions and men of large fortune'.[90] (One second-hand book dealer was advertising the same seven-volume 1851 edition, with index, cloth bound in 5 thick quarto volumes for just £8 12*s.* 6*d.* in May 1852).[91] A cheaper edition at 4 guineas (or 12 shillings a volume)[92] was issued in 1856, but though described as a second edition it appears to have been identical with the 1851 edition apart from the title pages.[93] Among the targeted market of the economy edition were Catholic clergy, and the publisher's prospectus issued in June 1855 appealed specifically to the desire of clergy 'to assist in preserving the native literature and language of Ireland'. The prospectus included approbations solicited from a large number of bishops. Dr David Moriarty, bishop of Kerry, praised the 'truly magnificent edition' of the Annals that had been published some years earlier, observing that Hodges and Smith had 'preserved from oblivion, and transmitted to posterity, this glorious record of the history and language of our country'.[94] John MacHale, archbishop of Tuam, mentioned its educational potential. Dr Slattery, archbishop of Cashel, saw it as contributing to 'the preservation of the national language and literature of Ireland', while Dr John Ryan, bishop of Limerick, welcomed the fact that:

warned Windele not to tell the publishers that he had obtained some sheets from O'Donovan himself.

90 Circular letter from Hodges and Smith, June 1855 (RIA, 12 I 15, p. 571).

91 Advertisement of Thomas Connolly, of 10 Ormond Quay, in *Irish Ecclesiastical Journal*, 142 (May 1852), 79.

92 Circular letter from Hodges and Smith, June 1855 (RIA, 12 I 15, p. 571).

93 The 1856 edition has been twice reprinted in the twentieth century. In 1966, AMS Inc, New York, an American reprint company made the seven-volume set available once again, still aimed at the library market. New sets of this edition were retailing in Dublin for IR£370 in the early 1980s. A further lavish reprint in 1990, by Éamonn de Búrca, was again aimed at the specialist collector, the publisher once again adopting the nineteenth-century strategy of printing a formal prospectus, citing P.W. Joyce to the effect that 'it was the greatest and most important work ever issued by any Irish publisher', and emphasising that the new 1990 reprint would be 'printed in Ireland' (*The Annals of the Four Masters: prospectus by Eamonn de Burca* [Dublin, 1990].) This 'third' edition was being advertised for sale by the publisher in 2003 at €765 [IR£602.50] (*De Búrca Rare Books, catalogue 66,* Spring 2003 (Dublin, 2003) pp. 89–90).

94 Hodges and Smith prospectus [June 1855] (RIA, 12 I 15/20).

> the only work which gives a full, impartial, and authentic account of Irish affairs, *before* as well as after the English invasion, is at length to be made accessible to Irish students, in the Irish original, with a faithful English translation, learnedly illustrated – and that Irish publishers are the spirited originators of an enterprise as perilous as patriotic and useful.[95]

Indeed the general tenor of these approbations emphasized the value of bringing such a text into the public sphere, available not merely to the wealthy elite, but also to 'those whose straitened circumstances forbade them to become members of the Archaeological Society',[96] but who were nonetheless anxious that a work of such national importance should be widely disseminated.

In reality, most general readers would have had to settle for the snippets from the annals that John O'Donovan had used to illustrate his essays in the *Dublin Penny Journal* (1831–7), which could claim a circulation of 40,000 in its first two years, and later the *Irish Penny Journal* (1840–1).[97] Philip McDermott and Owen Connellan likewise used material from the annals in the late 1840s to illustrate their historical contributions to popular magazines and newspapers including the *Nation*.[98]

CRITICAL ACCLAIM

In academic and literary circles O'Donovan's edition of the Annals was highly acclaimed, and the publishers were not slow to use extracts from published reviews as advertising for the work. Extracts from favourable reviews published in the *Gentleman's Magazine*, the *Irish Quarterly Review*,

95 Ibid.

96 Ibid., approbation of Thomas Feeny (RIA, 12 I 15/20, p. 575).

97 On the background context of these periodicals see Barbara Hayley, 'A reading and thinking nation: periodicals as the voice of nineteenth-century Ireland', in Barbara Hayley and Enda McKay (eds), *300 years of Irish periodicals* (Gigginstown, 1987), pp. 29–48.

98 See newscuttings inserted in Philip McDermott's own copy of Connellan (ed.), *The annals of Ireland*, preserved in the library of TCD (shelf-mark: EPB, Press A.7.15).

the *Dublin University Magazine*, the *Ecclesiastical Journal*, and the *Dublin Review*, were circulated in 1855 together with an order form inviting subscriptions at 12 shillings a volume.[99] The praise for O'Donovan's work was of the highest order. The enthusiasm of the *Quarterly*'s reviewer was typical:

> Whether we regard the industry and impartiality of the original compilers, the immense learning and extensive researches of the editor, or the exquisite typography of the volumes, it must be admitted that these Annals, as edited by Dr John O'Donovan, form one of the most remarkable works yet produced on the history of any portion of the British Isles. The mass of information which they embody constitutes a collection of national records, the value of which can never be superseded ... Standing thus alone, it must maintain a high place among the great literary monuments of the world, so long as the study of history continues to retain the charms which it has ever possessed for men of cultivated and philosophic minds.[100]

Such was the importance attached to O'Donovan's Annals that the review published in the *Irish Ecclesiastical Journal* extended over four monthly issues in 1848. Drawing attention to the value of Irish annals in general as a historical source, the reviewer asserted that 'they are a wonder, and without parallel in the world; and appearing in a digested and uniform form in the Annals of the Four Masters, they invest that collection with their own antiquity, and identify their compilers with the eye-witnesses of the facts recorded'.[101]

The publisher protested, somewhat ruefully, that the publication of the Annals had not been embarked on for profit:

99 'Opinions of the Press' (RIA, 12 I 15, pp. 585–9). The 1856 price was thus 4 guineas for the seven-volume set.

100 Extract from *Irish Quarterly Review*, 1 (1851), cited in 'Opinions of the Press' (RIA, 12 I 15, p. 587).

101 Review of John O'Donovan (ed.), *Annals of the Kingdom of Ireland* 3 vols (Dublin, 1848), in *Irish Ecclesiastical Journal*, 5 (1848), 79.

> For ourselves we may say with confidence, that in bringing out the work, time and expense have been secondary considerations, but we must admit that the expenditure of both has been much more than we estimated when we commenced the undertaking.[102]

Perversely, the absence of financial profit from the venture served to enhance George Smith's status as a public-spirited entrepreneur who, at a time of severe economic depression, was prepared to 'make this work an honour to the nation rather than a profit to himself, to this end sparing neither pains nor expense so that he might secure for his favourite object all the excellencies which editorship and art could confer'.[103]

The *Irish Ecclesiastical Journal* reviewer went on to portray the work as an icon of Irish ability to triumph over adversity, 'In a year of famine and great mercantile depression appeared the work whose title heads this article, as it were an earnest of intellectual propriety, and an omen of national convalescence'.[104]

The very warm reception accorded the publication of the Hodges and Smith edition of the Annals in academic circles, both in terms of the quality of O'Donovan's work as editor and the prestige character of the finished product, cemented the reputation of the Annals of the Four Masters as one of the foremost primary sources for the history of Ireland from earliest times to the beginning of the seventeenth century. At the time of its publication there were conscious and unconscious echoes of the original labours of the Four Masters in the 1630s, the most persistent (and most misleading) of which has been the notion that the work of the Four Masters was conceived as an antiquarian rescue mission in the 1630s. It was Petrie, O'Donovan, and their mid-nineteenth circle that were preoccupied with the idea of rescuing from oblivion the fragmentary primary sources of the Irish past, and they were inclined to project their antiquarian concerns onto the original annalists at work in an earlier era of upheaval and social change in the 1630s. O'Donovan's

102 Circular letter from Hodges and Smith, July 1851 (RIA, 23 I 15, p. 300).

103 Review of O'Donovan (ed.), *Annals of the Kingdom of Ireland*, in *Irish Ecclesiastical Journal*, 5 (1848), 123.

104 Ibid., (1848), 123.

recording of the fact that 'the translation was commenced in January in the year 1832, and finished on the 3rd of December 1833',[105] was probably a conscious echo of the Four Masters statement of three centuries earlier to the effect that they commenced the annals in 1632 and completed it on 10 August 1636.[106] O'Donovan's dedication of his edition similarly made explicit reference to Míchéal Ó Cléirigh's original dedication of the Annals to their 1630s patron, Fearghal Ó Gadhra, and he thanked those who had 'eminently distinguished themselves by their exertions in promoting the study of Irish History and Antiquities' for promoting 'the cause of ancient Irish literature, at a period when it had fallen into almost utter neglect'.[107]

Although O'Donovan's achievement in editing the Annals of the Four Masters has remained unsurpassed, it must be noted that his edition of the pre-1171 material has long been regarded as unsatisfactory. As Eugene O'Curry was careful to point out in his *Lectures on the manuscript materials of ancient Irish history*, O'Donovan did not have access to either of the known autograph manuscripts of the annals for the pre-1171 period, and hence was not in a position to produce a fully accurate edition. O'Curry regarded it merely as the standard edition 'in the present state of our knowledge' in 1861.[108] Despite this caveat O'Curry was generous in his praise of O'Donovan's scholarship, but reserved his highest praise for the publisher, George Smith.

> I cannot pass from the subject of this lecture without recording the grateful sense which I am sure all of you ... must feel, as I do, of the singular public spirit of Mr George Smith, at whose sole risk and expense this vast publication was undertaken and

105 Quoted in a review of the 1848 edition published in the *Gentleman's Magazine*, n.s. 20 (1848), 385, and referring only to the translation of the post 1171 section of the text.

106 O'Donovan (ed.), *Annals of the Kingdom of Ireland*, i, pp. lxiv-lxvii.

107 Ibid., pp. v-vi.

108 Neither O'Curry nor O'Donovan were aware, for instance, that the exemplars they used as the basis of the text and translation had been heavily abridged (Bernadette Cunningham, 'The making of the Annals of the Four Masters', University College Dublin, PhD thesis, 2005, chapter four).

> completed. There is no instance that I know of, in any country, of a work so vast being undertaken, much less of any completed in a style so perfect and so beautiful, by the enterprise of a private publisher. Mr Smith's edition of the Annals was brought out in a way worthy of a great national work, nay, worthy of it, had it been undertaken at the public cost of a great, rich, and powerful people, as alone such works have been undertaken in other countries. And the example of so much spirit in an Irish publisher – the printing of such a book in a city like Dublin, so long shorn of metropolitan wealth as well as honours – cannot fail to redound abroad to the credit of the whole country, as well as to that of our enterprising fellow-citizen.[109]

O'Curry was here echoing the sentiments of the many of his contemporaries who saw the publication of these volumes as transcending the mere academic achievement of producing a scholarly edition of an important primary source for the history of Ireland. As one reviewer had observed in 1851:

> At a period of unexampled commercial prostration and disaster, and when, especially in Ireland, the social system was shaken to its foundation ... they [Hodges and Smith] have again come forward to demand national gratitude by the publication of the greatest original work which has ever issued from the Irish press.[110]

Emerging as a beacon of hope from the gloom of famine and political unrest in 1840s Ireland, the publication of the Irish text and translation of the Annals of the Four Masters in a deluxe edition brought the Gaelic past unequivocally into the public sphere, and was welcomed by all as a tangible expression of patriotic endeavour and a monument to the resilience of the human spirit.

109 Eugene O'Curry, *Lectures on the manuscript materials of ancient Irish history* (Dublin, 1861), p. 161.
110 *Irish Quarterly Review*, 1 (1851), p. 697.

Historiography, heritage, inheritance

Irish castellology and Leask's Irish castles

TADHG O'KEEFFE

THE PUBLICATION AT THE END of the twentieth century of substantial books on Irish castles by Tom McNeill, a lecturer in archaeology in Queen's University Belfast, and David Sweetman, then senior archaeologist with Dúchas, the Irish State's heritage service, both established and respected authorities in the field of castle-studies or castellology, represented a bonanza for students of castles in Ireland, and indeed of the Middle Ages in Ireland. In bringing closure to a century of castle-research, these books represented an extended moment of maturation for Irish castellology, a time when the volume of completed empirical research had reached the threshold at which the stock-taking exercise of synthesis was warranted.

More than half a century separates these *fin-de-siècle* syntheses from the first edition of the most famous and widely read book on Irish castles, *Irish castles and castellated houses* by Harold Leask (1882–1964), published in 1941 by Dundalgan Press, the Dundalk-based company founded in 1859 by William Tempest (Fig. 1). Leask's *Castles*, or 'Leask' as Irish castellologists invariably identify it, is a classic of Irish historical scholarship. Reprinted by the same publisher seven times (1944; 1946; 1951; 1964; 1973; 1999; 2003), with three of those editions being revisions (1944; 1946; 1951) that were faithful to the appearance of the original, the book

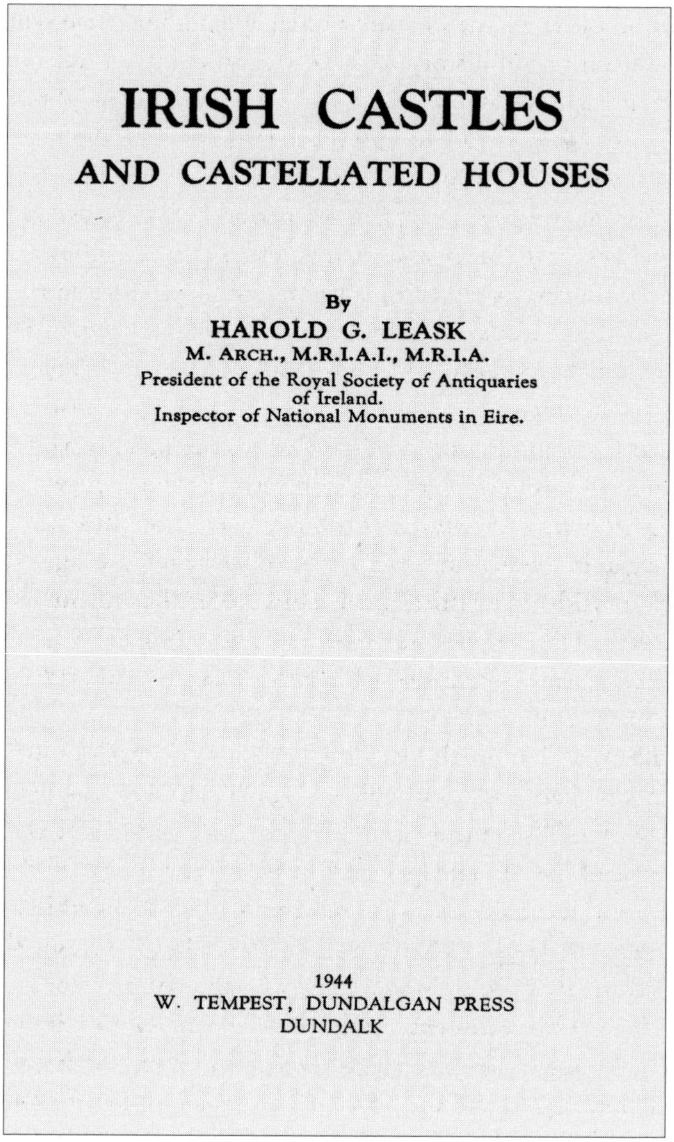
IRISH CASTLES
AND CASTELLATED HOUSES

By
HAROLD G. LEASK
M. ARCH., M.R.I.A.I., M.R.I.A.
President of the Royal Society of Antiquaries of Ireland.
Inspector of National Monuments in Eire.

1944
W. TEMPEST, DUNDALGAN PRESS
DUNDALK

1 Title page of *Irish castles*, first revised edition

remains in print. As the first substantial and authoritative synthesis of the architecture and history of 'the castle' in medieval Ireland, it was, prior to the late 1990s, the principal (and usually the only) port of call for archaeologists, historians, historical geographers and others wishing to find some general information about castle architecture. Other syntheses had of course appeared in the period between 1951 and 1997 – Mike Salter's small *Castles and stronghouses of Ireland*, published in 1993 under the unfortunately-titled imprint of Folly Publications, deserves mention – but Leask's vision informed every one of them.

Between the covers of this book Leask served up a simple model of castle development, based on accounts of individual castles that are arranged typologically and synchronized chronologically with a simplified understanding of medieval Irish history. The same writer's *Irish churches and monastic buildings* triology (1955–60), also published by Dundalgan Press, has had comparable historiographic significance in the field of Irish ecclesiastical architecture, being the only survey of such chronological scope and sustained interpretative originality to have been published in the century following Arthur Champneys' *Irish ecclesiastical architecture* of 1910.

Irish castles is a much-loved book among bibliophiles. Its 'analog' character is charming in this 'digital' age, its line-drawing illustrations – the work of the author himself – are both models of clarity and reminders of the draughtsperson's art before the arrival of Photoshop (Fig. 2), and its content is as jargon-free as is possible within a specialized field. Castellologists have a particular affection for it, since it stood alone on the shelf for so long. If I may be allowed a personal reminiscence, if only for fellow castle-scholars who will have their own recollections, I bought the 1973 reprint (which had a red, with black interweave, cloth cover, and the familiar full-page image of Rathmacknee Castle on the dust-jacket) about 1975, and later transported it on the bicycle carrier when, as a teenager, I pedalled around parts of the country looking for and at castles. The castles that featured in the main body of the book had *de facto* importance and therefore had to be visited; castles that featured in his appendix of castles, listed county-by-county, had an aura of significance by virtue of being included, even though he usu-

2 Leask's drawings of three internal features of Irish castles of different dates, from *Irish castles*, chapter x

ally said little or nothing about them, and they demanded to be visited as well. Every now and again I would come upon a castle that I thought merited inclusion in 'Leask', as if that was the final goal of every castle-builder and the ultimate accolade that one could bestow on a castle. I had no conception of a research universe outside 'Leask', or any sense of a system of evaluation that did not have Leask's book at its centre. The point at which I could visit castles and not bring 'Leask' with me was the point at which I regarded myself as having successfully negotiated some rite of passage, from interested amateur to serious student, even a specialist.

More than sixty years after its first appearance, and more than fifty since its last revision, Leask's *Irish castles* can no longer be as valued for its factual or interpretative content. In my capacity as a teacher, I started to exclude it from student reading lists after 1997, when McNeill's book appeared, and removed it completely after 1999, when Sweetman's book was published. I now recommend McNeill's book as a general course text on the grounds that it sells a far more convincing 'story', and I recommend Sweetman's book as a surrogate for Leask's book on the grounds that it balances a traditional, 'Leaskian', perspective with up-to-date factual information.

SOME NOTES ON IRISH CASTELLOLOGY BEFORE 'IRISH CASTLES'

Despite his exalted position in the history of Irish castellology, Leask was not the first modern scholar to write about Irish castles, either as individual castles or as variously constituted collectives of castles. Indeed, the pre-1941 historiography of Ireland's castles is festooned with famous names, from John Henry Parker, the influential student of English medieval architecture whose contribution to Irish castellology (Parker 1860) has been discussed by Colm Donnelly (1995), to Vere Gordon Childe, one of the twentieth century's most celebrated prehistorians, who investigated a 'castle mound' at Doonmore in Antrim (Childe 1938).

Ireland's modern tradition of castellology was properly inaugurated, however, in the first decade of the twentieth century with the publications of two great scholars, Goddard Henry Orpen, the celebrated historian of Anglo-Norman Ireland, and Thomas Johnson Westropp, the prolific Limerick antiquary (see Duffy 2001 and Ashe FitzGerald 2000 for respective biographies).

Orpen and Westropp pursued independent interests in the field. The former wrote detailed histories of several Irish castles and houses, among them Athlone and Newcastle McKynegan castles, both thirteenth century and later (1907a; 1908), while his seminal four-volume historical survey of Anglo-Norman Ireland (1911–20) acknowledged *inter alia* the importance of castle-form to the understanding of that period of history. Westropp, the most prolific writer among the numerous Irish antiquaries of his age (although the Kildare-born Lord Walter Fitzgerald runs him close), wrote about individual castles, such as Desmond Castle in Newcastle West in his native county of Limerick (1909), and published important surveys of castles in Clare and Limerick (1900; 1907). The defining event in this early phase of Irish castellology, and indeed the defining event in the history of Irish castellology, was the dispute between these two men on the chronology and cultural context of the type of earthen mound known to their generation as the 'mote' and to our generation as the 'motte' (see Herity 1994 for a revival of 'mote'). Briefly, Westropp argued that these mounds were native Irish monuments of pre-Anglo-Norman date (1904; 1906), while Orpen counter-claimed them to be the remains of Anglo-Norman castles (1906; 1907b). These were not independent enquiries: each was aware of the other's writings. Orpen's view prevailed, thanks in part to his deployment of historical evidence. With the benefit of scholarly hindsight we now recognize that Orpen and Westropp were not always discussing the same type of monument – Westropp's 'motes' constitute a more morphologically heterogenous group than Orpen's – so that we can hardly characterize their debate as a straight gun-fight over a single monument type. Nonetheless, so convincing was Orpen's argument about these monuments that many years later Leask could acknowledge the historian as having done 'his share' for Irish castellology by having 'drawn attention to the earthen fortresses of the Anglo-Norman invaders', many of which, 'in the then state of our knowledge, had been regarded as native works of prehistoric date' (1951, 2).

Orpen's opposition to Westropp's interpretation parallels that of Ella Armitage to G.T. Clark's interpretation of England's motte-and-bailey castles as works of the (Anglo-Saxon) English (Clark 1884–5). This was no coincidence. Armitage was, with Clark and John Horace Round, a member of the triumvirate of pioneering castle researchers in Victorian-era England (Counihan 1985; see also Coulson 1996). She and Orpen corresponded (Counihan 1997), and the Orpenesque interpretation of earthwork castles featured in her own classic survey of 'early' castles in these islands (Armitage 1912). Incidentally, Armitage was one of the few women to have written about castles in Britain or Ireland over the past century, but for Ireland she was pipped to the post by Constance Adams, whose book on the 'ancient' castles of Ireland and their 'fortress histories and legends' (1904) is of historiographic interest, even if its content is as lacking in discriminaton as its title suggests.

LEASK'S 'STORY' AND ITS MODIFICATIONS

English castellologist Matthew Johnson used the phrase 'the story' to describe the consensus view, and thus the dominant narrative, of the development of castle architecture in England (2002). That term has been appropriated by others (O'Keeffe 2001; Liddiard 2004). It need hardly be repeated here that Leask's version of 'the story' was the only version that existed for Ireland for many decades.

We will not dwell on the differences between the formulation that appeared in the 1941 edition and that which appeared in the last revised edition of 1951, but will treat the latter as his 'mature' statement, made before he turned his attention to a more systematic study of ecclesiastical architecture. Here is an outline of his version of 'the story', glossed with comments on the views of McNeill and Sweetman on these very points.

Defining 'castle'

The first point to be made about Leask's 'story' is that he had a rather narrow understanding of what 'castle' connoted in the Middle Ages. Acknowledging its broad definition in the modern world, he insisted that it signified in its 'strictest' medieval usage a 'fortress of mortared stone erected by a king or noble to guard a strategic position, or overawe a district, and serve as a residence for the owner or his deputy and retainers' (1951, 5). He presented a survey of such features of Irish castles as drawbridges and machicolations, as well as of such siege engines and weapons as trebuchets, mangons and cross-bows as described and illustrated (Fig. 3) for France by Eugène Emmanuel Viollet-le-Duc, to reinforce his view of the importance of defensibility in the definition.

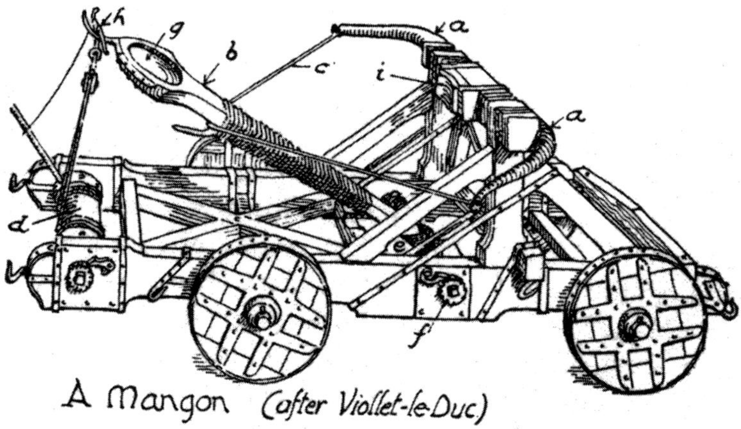

3 Leask's drawing of a mangon, from *Irish castles*, chapter III

Underpinning *Irish castles* thereafter is the assumption, never quite made explicit, that 'the castle', both in Ireland and more universally, had evolved in a Darwinian sense as a military machine between the twelfth century and the sixteenth. The actual argument for a process of natural selection of castle design that was responsive to the changing nature of seige warfare had been made some decades earlier (Hamilton Thompson 1912).

This matter of a proper definition of 'castle' is central to the entire castellological tradition of Europe. Leask simply articulated the view, first articulated with real force in the English language (to my knowledge) by G.T. Clark (1884–5), that a castle is fundamentally a military object. McNeill generally avoids definition in his book, but Sweetman offers a definition similar to Leask's – 'the castle is essentially feudal and is the fortified residence of a lord in a society dominated by the military' – and, like Leask, accompanies this with illustrations of siege machines (1999, 41). The simple understanding of the concept of 'castle' as purveyed by Leask and Sweetman for an Irish context has already been the subject of critique (O'Keeffe 2001; 2004), and Charles Coulson's recent discussion of this issue with respect to Ireland, Britain and France

(Coulson 2003) underscores with ample exemplification the foolishness of assuming 'castle' to be a simple concept.

Earth-and-timber castles

Leask included earthen fortresses of 'mote' and 'mote-and-bailey' type within the rubric of 'castle'. Leask saw these earth-and-timber castles, as we would now describe them, as 'progenitors' of stone fortresses, 'ill-suited to be permanent residences' (1951, 5, 25), rather than as fully-fledged castles in their own right. The two later syntheses offer a very different perspective on this matter. Both McNeill and Sweetman emphasize the importance and longevity of these earth-and-timber castles, with the latter writer devoting considerable space to discussions of the two principal types, one of which, the so-called 'ringwork castle', was not known to Leask. Indeed, one difference between McNeill and Sweetman is that the former is far more sceptical about ringwork identifications than the latter, and therefore has less to say about this class of monument.

In common with many mid-century historians and castellologists, Leask understood the association of 'castle' and 'feudalism' to be immutable, and so he excluded such native Irish earthworks as ringforts from the category, asserting them to be products of 'a different outlook on life, of different manners and customs in both war and peace' (1951, 6). He did acknowledge that some high-status native Irish structures of pre-1169 date were recorded by contemporary annalists as 'castles', but he concluded that they were of perishable material and he did not regard them as comparable with the castles of the Anglo-Normans. While his view of ringforts is still generally held, albeit with some qualification, the pre-Norman castles of the annals are now regarded, as we will see below, as having been of motte form (Graham 1988; O'Keeffe 1998).

It is interesting to note that Leask wrote *Irish castles* in a pre-excavation age, and had no excavated data of any significance to draw on. After about 1950, however, a programme of excavation and survey of medieval sites in Northern Ireland, particularly in County Down, threw some light on the structural histories of stone castles and illuminated especially well the complex histories of the Anglo-Norman earthworks.

Dudley Waterman and Martyn Jope, both medievalists, and Pat Collins, a prehistorian, were the principal figures, but neither Waterman nor Jope produced a synthesis, although the appropriate sections of the magnificent *Archaeological survey of Co. Down*, published under Jope's editorship in 1966, the year after Leask's death, does give some impression of what might have issued from this Northern enterprise. Leask's book, then, remained 'the text' as far as the wider public was concerned, and so the new data coming out of Belfast remained largely unfamiliar to those unaware of the existence of specialist papers in the *Ulster Journal of Archaeology*. One could argue that Leask's characterization of earthwork castles as 'progenitors' (1951, 5) of stone castles may have deterred him anyway from developing the full implications of the evidence from eastern Ulster.

It is worth straying from Leask for a moment to make some general observations on the study of earth-and-timber castles in the second half of the twentieth century. That Northern Ireland tradition of investigating mottes and motte-and-bailey castles never actually extended southwards, so the number of investigated monuments of these types in the Republic is a fraction of that north of the border, although the number of actual sites in the Republic is greater. McNeill has suggested that the explicitly English character of these earthworks as revealed by Orpen, himself a purveyor of 'overt Unionism', may have rendered these undesirable subjects of analysis, both in the Free State and in the Republic that succeeded it (1997, 2). He may have a point: as late as 1970 Athlone Castle was threatened with destruction on account of its colonial, English, associations (O'Keeffe 2004, 139). Ironically, the evidence that emerged from excavations at earthwork sites like Rathmullen (1981–2), and from the combination of historical and archaeological research at sites like Duneight (Waterman 1963), suggests that the pre-Norman Gaelic Irish of eastern Ulster were actually castle-builders in all but name, and that Orpen's (and Leask's) black-and-white view of castle-building as the cultural perogative of the newly-arrived English requires modification. The Ulster research could, in other words, be used to feed a nationalist agenda. It has not, and we should be thankful that so explicit a political project has been eschewed, but the irony is inescapable.

Indeed, the past half-century has, to my mind, seen southern Irish castellology develop a stronger engagement with 'the colonial' than with 'the native', a view that can be defended by reference to the archaeology of the earthwork castle. On the one hand, there has been a general disregard of the very issue of native incastellation, particularly of the pre-1169 era. In the course of the past half-century English castellologists have rejected that simple equation of Conquest = Feudalism = Castle that Armitage had argued for in 1912, and have pursued an active interest in pre-Conquest incastellation, either by Anglo-Saxon lords or by Normans resident in the England of Edward the Confessor (Saunders 1977). But Irish castellologists have barely developed a parallel paradigm for this island. Can we blame Leask? No. He had mentioned the annalistic references to native Irish incastellation in the twelfth century, even if he was rather sceptical about the accuracy of the annalists' terminology and about the significance of whatever structures they were describing as castles. It is ironic, though, that these stray references to castles came to be seen in the 1970s as part of a whole register of indicators of an indigenous, pre-Norman, feudal tradition (see Ó Corráin 1978, for example), while the same decade saw greater archaeological interest in the colonial incastellation, as witnessed by the 'discovery' of the ringwork castle (Barry 1983).

Anglo-Norman stone castles

Leask identified the great period of castle-building as running from *c.*1180 to *c.*1310, the former date chosen on the understanding that building in stone was not possible until the new lordship established in 1169–71 was properly secured, and the latter date chosen on the grounds that social and political conditions would have militated against much building activity as soon as the fourteenth century was underway. His account in *Irish castles* of castle-building during this long period of thirteen decades was very close in content to a paper he had published in the *Archaeological Journal* in 1937.

Leask identified castles of the first half of this period as invariably having keeps, great towers that he understood to have been potential final refuges if castles were attacked. He described rectangular examples

4 Leask's drawing of Carrickfergus, from *Irish castles*, chapter III

that stand flush with the enclosing walls, as at Carrickfergus (Fig. 4), or are isolated within the walls, as at Trim (Fig. 5), round and polygonal examples, which are a little less common, and so-called 'towered' or 'turreted' examples, such as Carlow, Lea and Ferns which he identified as peculiar to Ireland in the thirteenth century. He described the castles of the second half of this period as generally 'keepless', citing castles like Roscommon and Ballymote as typical examples, although he conceded that the keepless castles in Dublin, Limerick and Kilkenny actually belonged to the early 1200s. He explained that the keep had become obsolete in the later 1200s once greater emphasis was placed on enclosing walls, tower-form gatebuildings, and flanking towers.

Now, Sweetman's account of castle-building during this period follows a slightly different pathway, and gives a prominent place to the so-called hall-house, another newly identified type, but it follows Leask on many major points, such as the nature and chronology of the 'towered keeps', as well as in respect of a lack of detailed cross-references to contemporary English and Welsh castles. It could be argued, of course, that the sequence of buildings is well known, and that the type of synthesis that marks the Leask-Sweetman axis has validity, but McNeill's commentary on the same basic corpus is more sensitive to the factors that influenced design, as witness his grouping of castles under the headings 'English fashion' and 'divergent tradition'. One criticism of all these syn-

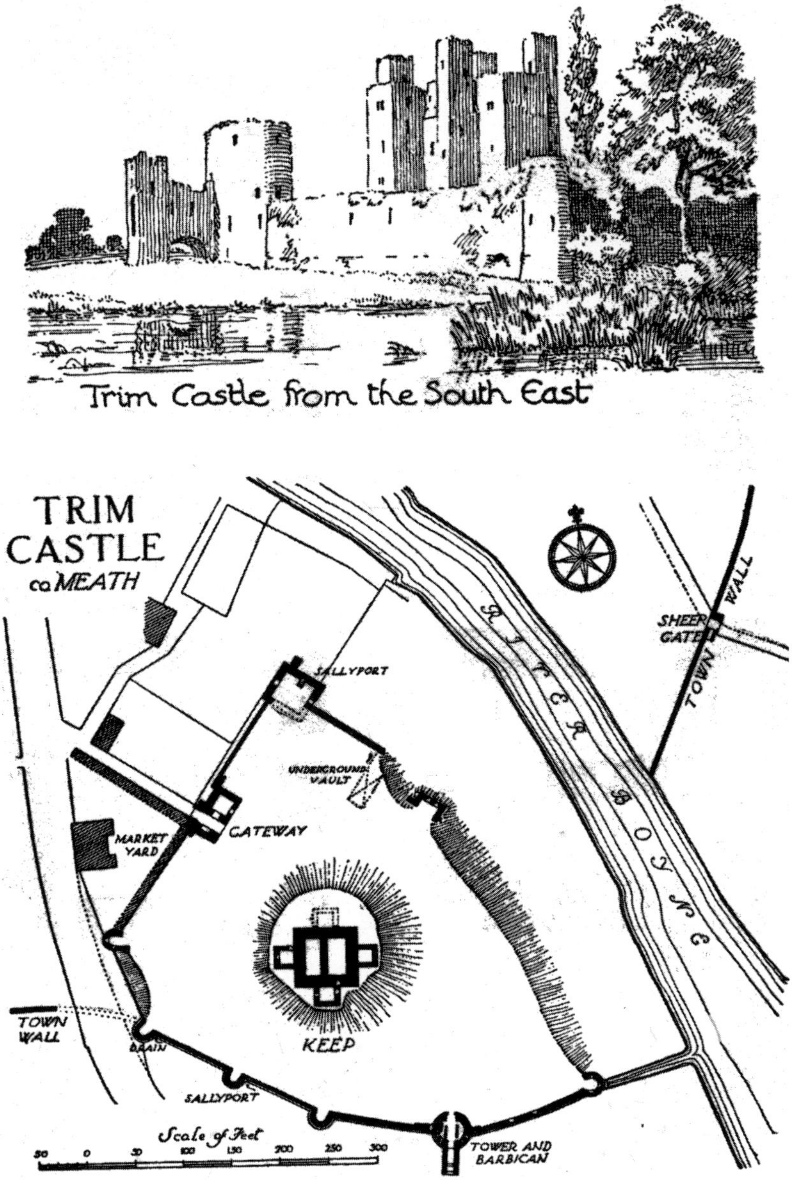

5 Leask's drawings of Trim, from *Irish castles*, chapter III

theses is that they pay too little attention to symbolic defence, such as the signification of defensibility by the mere presence of crenellations (see Coulson 1979; 1982). The inherent symbolism of the design certainly merits closer attention for the great towers or 'keeps': the rather anachronistic-looking designs (and Romanesque details) of late twelfth-century towers at places like Carrickfergus, for example, may have been motivated by a desire among patrons to historicize architecturally their new Irish landholdings, while the unusual plan of the great tower at Trim may have been based on the plans of Orford and Dover, two English royal castles, with the explanation to be found in Hugh de Lacy's close relationship with Henry II.

The tower-house tradition

Moving to the fourteenth century, Leask identified a hiatus of building activity during most of the 1300s, and he attributed it to a combination of war and famine. He noted that the pace of building, of churches as well as of castles, picked up again in the early fifteenth century and that it remained brisk for the following two hundred years. Sweetman shares this view of a fourteenth-century hiatus, but McNeill requests more careful reflection, drawing attention, for example, to the problem of actually dating castles. Leask correctly identified tower-houses – 'fortified houses, the ordinary and typical residences of the Irish and Anglo-Irish gentry' (1951, 76) – as the typical castles of the two centuries that followed the fourteenth, and, anticipating Ó Danachair's map almost thirty years later (1979), he noted their heavy concentrations in rich farmlands that stretch across southern Ireland from Limerick to Wexford. Leask's interpretation of the tower-houses as essentially fortified residences has endured, though McNeill is less supportive of this interpretation than Sweetman, pointing out that many allegedly defensive features had little practical use and were probably display-related.

Discussing tower-house origins, Leask attributed great importance to the acts of parliament of 1428–9 and 1449 that offered £10 subsidies to those loyal servants of the crown within the lands of the Pale who would build castles of a specific size. Believing no tower-houses to predate *c.*1400 at the earliest, he saw these inducements as likely stimuli for the generation of the small tower-form castle type across Ireland. This explanation of tower-house origins has generated much debate, but recent years have seen rejections of it. McNeill, needless to say, regards the use of the so-called '£10 castle acts' for the purpose of establishing tower-house chronology as 'unwise'. My own view, for what its worth, is that the money on offer was insufficient to effect any new architectural development – it was too small an amount to influence intending castle-builders to change their plans, or to allow those not already on the property ladder to start building – and that its granting was more a contract of loyalty between the government of the crown and the Pale landowner than a practical aid to the incastellation of the Pale (O'Keeffe 2001). Sweetman, however, aligns himself with Leask on this matter, but is even more confident of the importance of the 1428–9 subsidy. He describes it as 'an inescapable historical fact about the origins of the tower house' and claims that it 'provided the impetus to build the typical tower house of the eastern half of Ireland' (1999, 137).

Plantation-era castles

The latest-dated tower-house mentioned by Leask is Derryhivenny near Portumna, County Galway, built in 1643. By that date, new 'many-windowed and gabled' houses retaining 'sufficient defensive features to justify, in some degree, the designation castle', had appeared in Ireland. Leask identified the Plantation landscapes of southern Ireland as home to the most interesting array of types in the late 1500s and early 1600s, with some buildings aspiring to Renaissance grandeur while retaining their defensibility (Mallow Castle, for example). He also recognized the great building boom in Plantation Ulster in the period 1610–20. He saw the medieval castle-building tradition as all but over by 1640, when new artillery-repellent structures marked the start of a new, post-castle, era of fortification. Sweetman distinguishes between 'fortified houses' and 'stronghouses' in this later sixteenth- and seventeenth-century period, and stresses the importance of fortification even more than Leask had done. But the designation 'castle' that is attached to many of these buildings is easily misunderstood as an expression of their defensibility, which is a point readily understood if one takes a wider, northern European perspective (Ronnes 2004). McNeill terminates his survey before this period.

WAS LEASK A THEORETICIAN OR A TYPOLOGIST?

Although the corpus of castles in Ireland differs significantly from that of Britain, Leask's 'story' matched that generated contemporaneously for British and European castles by scholars such as Sydney Toy (1955), insofar as 'castle' was understood to be a fundamentally military concept, and its dynamic of architectural design development was understood to be a search for an optimal balance between defensibility and domesticity according to political circumstances. However, it would be incorrect to regard Leask's book as a closely-argued example of such cause-and-effect archaeological thinking. He had a stronger allegiance to the concept of typology than he had a desire to explain dynamic process.

His chapters entitled 'Castles in towns' and 'Castles on rocks' give the game away. The buildings he discussed here, such as Roscrea, Castleroche, and Dunamase, could easily have been given homes alongside other castles in other parts of the book had he opted for a more explicitly interpretative narrative, had he devoted more mental energy to its plot, and indeed had he paid more attention to the English and Welsh comparanda with which he must have had some familiarity through his professional training. Instead, however, he grouped these castles together according to siting and location, factors that had little apparent bearing on their architecture.

The chapters on tower-houses are similarly revealing. After a thoughtful – though misjudged – discussion of the origins of the tower-house tradition, *Irish castles* descends into an account of individual buildings, dispersed somewhat illogically across several chapters, with a chapter on features of the towers wedged between them. Regional variation, perhaps the key to understanding the corpus, is barely acknowledged, although its importance should have been apparent to him from the castles that he saw and recorded.

Heretical though it may be, I would suggest that, for all his strengths as a 'reader' and recorder of complex structures, a skill on display in many of the individual studies of papers (listed by Lucas 1966), Leask was only moderately equipped for the challenge of interpretative synthesis that was posed by the corpus of castles that he studied. Yes, the fact that his two volumes on post-1170 ecclesiastical architecture were conceptually more advanced than *Irish castles* suggests that he had the capacity for a great, interpretative, synthesis, but it is appropriate to note that these volumes are actually more indebted to Arthur Champneys and his *Irish ecclesiastical architecture* on 1910 than he admitted.

Is Irish castles *yesterday's book?*

The fact that I choose McNeill over Sweetman as the book to replace Leask should not be of special interest outside of the classroom, but I wish to draw attention to there being a choice, and to advance the observation that the choice is not so much between books but between ways of seeing. This is an important issue. Castles are of shared interest to scholars of many disciplines, so the epistemology of castellology really matters. Castles are also heritage objects, and are used in what might be described as 'cultural education', so here too the epistemology of castellology matters (for a discussion of this based on the controversy over Carrickmines Castle see O'Keeffe 2004). At stake is something more important than the destination of royalties when students are advised to make purchases.

Now, McNeill certainly effected the long-overdue radical break with the 'Leaskian' tradition in Ireland, not just with respect to the interpretation of individual castles or castle-types but with respect to the entire passage of castle-building history. He ignores the assumption of Darwinian evolution that implicitly underpinned *Irish castles* as well as other mid-century books on castles, and he eschewed the simple typologizing in which Leask engaged. Instead, his analysis of individual castles and groups of castles is built on a cognisance of castle-building traditions in Britain and beyond, a failing of Leask in *Irish castles*, as well as on a more sophisticated understanding of the geography and politics of lordship. My abandonment of *Irish castles* as student reading material is thus explained.

However, McNeill's synthesis notwithstanding, Leask's *Irish castles* cannot be characterized as 'yesterday's book'. Although he died in 1964 (see Lucas 1966) after a distinguished career as an inspector of monuments with the Commissioners of Public Works, his voice continues to be heard. His figure looms large over the choice between McNeill and Sweetman, and indeed between them and a third book, as yet unwritten for Ireland, that could develop a more explicitly cultural-theoretical understanding of 'the castle'.

There are two reasons why Leask is still part of our intellectual landscape. The first, I contend, is that David Sweetman's book (1999) raised the 'Leaskian' perspective from the grave to which McNeill had freshly consigned it, and garlanded it with the imprimatur of the state's archaeological service. Nowhere is this more apparent than in the county inventories of archaeological sites published by the Archaeological Survey of Ireland: the published descriptions of individual castles are straightforward, but the introductory sections to the *Castles* chapters reveal a clear debt to Leask, sometimes even down to the use of the word 'keep' rather than 'great tower' as is now normally preferred. Sweetman's book was not a rewriting of Leask; he admits that his original ambition was to 'update and upgrade' *Irish castles* but that he came to see this as neither 'practical nor desirable' (1999, i). Yet, while Sweetman's book is a new work, complete with long discussions of ringwork castles and hall-houses, two monument types that were unknown to Leask, its narrative style and its chronologically-ordered typology are clearly indebted to *Irish castles*. Members of its fan-base see it as Leask's successor, and venerate it as if McNeill's book had not appeared in the interim: the following endorsement on the dust-jacket by Terry Barry, another leading Irish castellologist, makes the point about genealogy very well: 'This is the definitive book on the archaeology of Irish castles. All of us who are interested in the subject have been *waiting for many years* for this up-to-date book which now undoubtedly *replaces Leask's classic work* which was written over half a century ago' (emphasis added).

The second reason is that Leask's position as doyen of Irish castellology was unopposed for so long that his views naturally filtered into the intellectual groundwater. Sheer familiarity with his work has ensured that his 'way of seeing', to borrow John Berger's phrase (1972), has come to be regarded by many as the only way. Indeed, that way of seeing had for many years informed choices of castles to be privileged with the protection afforded by National Monument status, and this in turn nourished the perception that Leask had worked out what these castles were really about. Of course, the development of alternative ways of reading castle architecture has not been affected by the continued presence of Leask's narrow, typological, vision; one can simply ignore the Leask view and happily argue that, for example, the use of the tower-house as a type by Ireland's different politico-ethnic communities – Gaelic-Irish, Anglo-Irish, and English – reflects a later medieval identification of architecture as an inter-cultural (O'Keeffe 2001) and intra-cultural (Tierney 2005) discursive tool. But whereas cultural-theoretical readings of comparable nature have now become commonplace in Britain (see Johnson 2002, for example), in Ireland these ways of seeing and thinking are still perceived as being on the margins, victims of a resistance to theory and a fetishization of 'data' that finds an eloquent but ignorant articulation in David Sweetman's attack on 'armchair' archaeologists who 'pontificate and produce marvellous theories without ever having done the necessary fieldwork to establish a database from which to draw their theories' (2003, 247). Our understanding of our corpus of architecture will be narrower in scope than it can be if we abandon the belief that there are ways other than the typological of understanding castle architecture. But the reconfiguration of Leask's half-century-old classic as *The medieval castles of Ireland* by David Sweetman might be interpreted more generally as evidence of a dogged conservatism in Irish castellology, and maybe even in Irish medieval studies in general.

BIBLIOGRAPHY

Adams, C.L. 1904 *Castles of Ireland: some fortress histories and legends*. London.
Armitage, E.S. 1912 *The early Norman castles of the British Isles*. London.
Ashe FitzGerald, M. 2000 *Thomas Johnson Westropp (1860–1922), an Irish antiquary*. Dublin.

Barry, T.B. 1983 'Anglo-Norman ringwork castles in Ireland: some evidence', in T. Reeves-Smyth and F. Hamond (eds) *Landscape archaeology in Ireland*. Oxford, pp. 295–314.
Berger, J. 1972 *Ways of seeing*. London.
Champneys. A. 1910 *Irish ecclesistical architecture*. London.
Childe, V.G. 1938 'Donmore, a castle mound near Fair Head, Co. Antrim', *Ulster Journal of Archaeology*, 1, 122–35.
Clark, G.T. 1884–85 *Medieval military architecture in England*. 2 vols. London.
Coulson, C. 1979 'Structural symbolism in medieval castle architecture', *Journal of the British Archaeological Association*, 132, 73–90.
Coulson, C. 1982 'Hierarchism in conventual crennelation: an essay in the sociology and metaphysics of fortification', *Medieval Archaeology*, 26, 69–100.
Coulson, C. 1996 'Cultural realities and reappraisals in English castle study', *Journal of Medieval History*, 22, 171–208.
Coulson, C.L.H. 2003. *Castles in medieval society. Fortresses in England, France and Ireland in the Central Middle Ages*. Oxford.
Counihan, J. 1985 'Mrs Ella Armitage, John Horace Round, G.T. Clark and early Norman castles', *Anglo-Norman Studies*, 8, 73–87.
Counihan, J. 1997 'Mrs Ella Armitage and Irish archaeology', *Anglo-Norman Studies*, 20, 59–67.
Donnelly, C. 1995 'John Henry Parker and his contribution to Irish Tower House studies', *Newsletter of the Group for the Study of Irish Historic Settlement* 5, 5–7.
Duffy, S. 2000 'Historical revisit: Goddard Henry Orpen, *Ireland under the Normans, 1169–1333* (1911–20), *Irish Historical Studies*, 37, 246–59; reprinted in the Four Courts Press edition of Orpen's work (2005).
Graham, B.J. 1988 'Timber and earthwork fortifications in western Ireland', *Medieval Archaeology*, 32, 110–29.
Hamilton Thompson, A. 1912 *Military architecture in England during the Middle Ages*. Oxford.
Herity, M. 1993 'Motes and mounds at royal sites in Ireland', *Journal of the Royal Society of Antiquaries of Ireland*, 123, 127–151.
Johnson, M. 2002 *Behind the castle gate: from Medieval to Renaissance*. London: Routledge.
Jope, E.M. (ed.) *An archaeological survey of Co. Down*. Belfast.
Leask, H.G. 1941 *Irish castles and castellated houses*. 1st ed. Dundalk.
Leask, H.G. 1937 'Irish castles, 1180 to 1310', *Archaeological Journal*, 93, 143–99.
Leask, H.G. 1955–60 *Irish churches and monastic buildings*. 3 vols. Dundalk.
Liddiard, R. 2004 *Castles in context*. Macclesfield.
Lucas, A.T. 1966 'Harold G. Leask, M.Arch., Litt. D., Past President' [and 'List of Published Works'], *Journal of the Royal Society of Antiquaries of Ireland*, 96, 1–6.
Lynn, C.J 1981–82 'The excavation of Rathmullen, a raised rath and motte in county Down', *Ulster Journal of Archaeology*, 44–5, 65–171.
McNeill, T. 1997 *Castles in Ireland. Feudal power in a Gaelic world*. London.
Ó Corráin, D. 1978 'Nationality and kingship in pre-Norman Ireland', in T.W.

Moody (ed.) *Nationality and the pursuit of national independence.* Belfast, pp. 1–35.
Ó Danachair, C. 1979 'Irish tower-houses', *Béaloideas*, 45–7, 158–63.
O'Keeffe, T. 1998 'The fortifications of western Ireland, AD 1100–1300, and their interpretation', *Journal of the Galway Archaeological and Historical Society,* 50, 184–200.
O'Keeffe, T. 2001 'Concepts of 'castle' and the construction of identity in medieval and post-medieval Ireland', *Irish Geography,* 34, 1, 69–88.
O'Keeffe, T. 2004 'Heritage, rhetoric, identity: critical reflections on the Carrickmines Castle controversy', in M. McCarthy (ed.), *Ireland's heritages: critical perspectives on memory and identity.* Aldershot, 139–51.
Orpen, G.H. 1907a 'Motes and Norman castles in Ireland', *Journal of the Royal Society of Antiquaries of Ireland*, 37, 123–52.
Orpen, G.H. 1907b 'Athlone Castle: its early history, with notes on some neighbouring castles', *Journal of the Royal Society of Antiquaries of Ireland,* 37, 257–76.
Orpen, G.H. 1908 'Novum Castrum McKynegan, Newcastle, Co. Wicklow', *Journal of the Royal Society of Antiquaries of Ireland,* 38, 126–40.
Orpen, G.H. 1911–20 *Ireland under the Normans.* 4 vols, Oxford.
Parker, J.H., 1860 'Observations on the ancient domestic architecture of Ireland', *Archaeologia*, 38, 149–76.
Ronnes, H., 2000 'The meanings of tower-houses'. MA thesis, University College Dublin.
Ronnes, H., 2004 'A solitary place of retreat: renaissance privacy and Irish architecture', *International Journal of Historical Archaeology*, 8 (2), 101–17.
Salter, M. 1993 *Castles and stronghouses of Ireland.* Malvern.
Saunders, A.D. 1977 'Five castles excavations: reports of the Institutes project into the origin of the castles of England', *Archaeological Journal*, 134, 1–156.
Sweetman, D. 1999 *The medieval castles of Ireland.* Cork: Collins Press.
Sweetman, D. 2003 'Archaeological survey: some sobering experiences', in J. Fenwick (ed.) *Lost and found: discovering Ireland's past.* Bray, 247–53
Tierney, A. 2005 'Castle-building and identity in Gaelic Ireland'. PhD thesis, University College Dublin.
Toy, S. 1955 *A history of fortification from 3000 BC to AD 1700.* London.
Waterman, D. 1963 'Excavations at Duneight, Co. Down', *Ulster Journal of Archaeology,* 26, 55–78.
Westropp, T.H. 1900 'Notes on the lesser castles or "peel towers" of the county Clare', *Proceedings of the Royal Irish Academy*, 21 C, 348–65.
Westropp, T.H. 1904 'Irish motes and early Norman castles', *Journal of the Royal Society of Antiquaries of Ireland*, 34, 313–45.
Westropp, T.H. 1905 'Irish motes and alleged Norman castles; notes on some recent contribution to their study', *Journal of the Royal Society of Antiquaries of Ireland,* 35, 402–6.
Westropp, T.H. 1907 'The ancient castles of the county Limerick', *Proceedings of the Royal Irish Academy*, 26 C, 55–108; 143–264.
Westropp, T.H. 1909 'The Desmond castle, Newcastle West, Co. Limerick', *Journal of the Royal Society of Antiquaries of Ireland*, 39, 42–58; 350–68.

The sovereign ghosts of Thomas Moore

HARRY WHITE

I BEGIN WITH Seamus Deane in *The Field Day anthology of Irish writing*:

> When read as the translation of something wild and savage into something regulated and polished, it [Moore's verse] becomes a metaphor of the translation of Catholic into Protestant, of native and antique authenticity into modern and equally native civility. Moore was the most outstanding figure in the long process of transformation that had begun in the eighteenth century and was to be carried on into the twentieth – the transformation of one culture into the idiom of another while attempting to preserve as much as possible of the old culture. It was a pitting of the foreign against the native, so that the foreign thereby would become Hibernicized and the native Anglicized. In other words, what had become a political problem for centuries was now becoming, self-consciously, a cultural issue. We may say that Moore's elegance was frequently reduced to enervation; we may say that his words do not live without their accompanying music, or that the music is violated by the artificiality of the words; we may laud or regret his influence on Irish nationalism. Yet, ultimately, the kind of issue raised by Moore's poetry, or, more par-

> ticularly by the *Irish melodies* and some of the *National airs*, lies at the heart of Irish writing for more than a century afterwards.[1]

The considerate nature of this point of view, and its considerations in respect of music in particular, provide a valuable corrective to the vertiginous descent that characterizes so unmistakably the reception history which attaches itself to Thomas Moore to the present day. By now, one feels, he has become something of a joke in the canons of Irish literature, or at best a superannuated figure lost to the ebullient modernism of the Celtic Revival and the magisterial confidence and achievement of Irish writing in the twentieth century. Emancipated as it most certainly is from the narrow claims of nationalism, Irish cultural history now consigns Moore to the oblivion of sentimentality, or at best deconstructs his work as a species of Victoriana, ultimately supervened by the more robust claims of those 'Irish classics' (the phrase is Declan Kiberd's) which followed him. It is not simply that Moore is absent from the imaginary museum of Irish writing (Kiberd's brilliant book typifies the general trend to overlook Moore almost completely), but that his place in Irish literary history is less and less secure, complicated on one side by the address which Moore himself makes upon the question of Irish music, and on the other by those canonic engagements with modernism which guaranteed to Yeats, Joyce and Synge an international reception that bypassed Moore or made him irrelevant.[2]

Almost from the outset, the difficulty of accommodating Moore's achievement into any kind of continuum or reception history was defined by the perception that he stole from the dull antiquarians a priceless trove of melody which was then burnished by his own poetic bravura and skill. Moore himself acknowledged that his interest in Irish music was stimulated by reading Edward Bunting's first collection of airs (published in 1797), but he was likewise alert to the political transformation

1 Seamus Deane, 'Thomas Moore (1779–1852)' in Seamus Deane (ed.), *The Field Day anthology of Irish writing*, 3 vols (Derry, 1991), i, pp. 1055–6

2 See Declan Kiberd, *Irish classics* (London, 2001). In this widely acclaimed and compelling reading of Irish literature, Moore's minor status is implicit in his absence from the pantheon of major Irish writers.

of expression which Bunting's collection clearly stimulated within him. That Moore himself would be overtaken by a more strident nationalism which added militant immediacy and subtracted the aesthetics of art music is part of the reception history which concerns me in this essay, but this supervention, together with Moore's subsequent projection as a hated darling of the Victorian parlour, should not be allowed to eclipse the fact that between 1808 and the advent of the Great Famine of the 1840s, Moore's voice became the conduit of a romanticized political sentiment which had its origins not only in Bunting and the modish appetite for Celticism, but in the United Irishmen, the failure of rebellion and the newly symbolic force of Irish music itself.[3]

The popularity of this sentiment can easily be identified as a primary agent in Moore's fall from grace. It might seem an exaggeration to suggest that the sheer ubiquity of the *Melodies* in particular, to say nothing of their iconic status in popular Irish culture, became something of an embarrassment to the avatars of the Celtic Revival, but there can be no doubt that Moore's achievement by the end of the nineteenth century was substantially reduced to a handful of songs which frequently became intermixed with the 'cloth eared' balladry of Thomas Davis (the phrase is borrowed from Seamus Deane) and the more general projection of Irish music as a pastime unworthy of serious scholarly attention. 'As for the peasant', Yeats remarked, 'he has his songs and his music'. He might have been thinking of Moore. By 1922, a reputation for cloying sentimentalism was reinforced by ridicule, as when Leopold Bloom passes the statue of Moore in College Street in *Ulysses*: 'He crossed under Tommy Moore's roguish finger. They did right to put him up over a urinal: meeting of the waters. Ought to be places for women. Running into cakeshops. Settle my hat straight. *There is not in this wide world a vallee.* Great song of Julia Morkan's ... Pupil of Michael Balfe's wasn't she?'[4]

I have never been sure of how strategic Bloom's stream of con-

3 For an account of Moore's understanding and reception of Irish music, see Harry White, *The keeper's recital: music and cultural history in Ireland, 1770–1970* (Cork, 1998), pp. 36–53.

4 James Joyce, *Ulysses* [1922] (London, 1992), pp. 205–6. Bloom is thinking of Moore's 'The meeting of the waters' and of Julia Morkan who features so prominently (as Aunt Julia, one of the 'Three Dublin Graces') in Joyce's 'The Dead'.

1 Thomas Moore, from *The poetical works of Thomas Moore collected by himself*, 10 vols (London, 1840–1)

sciousness is meant to be, but there can be little doubt that mocking Moore, when added to Bloom's incipient feminism (or rather femininity, to judge by what happens to him later in the novel), lends perspective to the kind of reception history I am trying to identify here. The join between Moore and Balfe also seems fortuitous, as does the unmistakable reference to 'The Dead', a narrative in which Irish music, if not Moore himself, is so structurally and emotionally decisive.[5] To think about the statue itself, as against Joyce's description of it in *Ulysses*, is to contemplate how very little has changed in the past eighty years (or more properly the past century, given 1904 as the year in which Ulysses is set) with regard to Moore's accommodation in the canons of Irish cultural history. Roguish finger or not, the urinal below him and the bird droppings which perpetually whiten his bronze pate ensure that a statue otherwise explicit in its intent of ennobling the Bard of Erin is thereby an image of some decrepitude and neglect. Unveiled in 1857, some five years

5 I have developed this argument in greater detail in my 'James Joyce, music and Dublin' in Harry White, *The progress of music in Ireland* (Dublin, 2005), pp. 111–20. Balfe's operas, notably *The Bohemian girl*, join the songs of Moore in Irish musical reception history as a code of popular remembrance and sentiment. This conjunction is explicit throughout Joyce's fiction.

after Moore's death, it is hard to conceive of the philistinism by which a monument dedicated to an artist is so utterly undone by the public lavatory built beneath it. But there it is: a statue which is so strongly suggestive of the arts as a central (rather than a peripheral) engagement in urban life becomes instead a permanent tribute to the dislocated condition of literature and music in Ireland throughout the nineteenth century. The rhetoric of statuary belongs and functions elsewhere, it would seem. In Vienna, or Bonn, or London. But not without irony in dear old Dublin.

This image of neglect (compounded, I freely admit, with instructive irony) calls to mind monuments of another kind and the lack thereof in respect of Moore. In an age suffused with multiple editions of writers and composers, to say nothing of the preservation and scholarly recension of even the furthest reaches of European literature and music, it is all the more striking to discover that as yet no complete works edition of Moore exists. The monumental nature of such an editorial enterprise would certainly justify the monument (that is, the edition) itself, but even more vitally it would restore to Moore's *Irish melodies* that context of which they are so sorely deprived, namely his other works. No-one could usefully deny the largely negative reception which history attaches to Moore, but few people, likewise, could deny the better access to Moore's world-view (I should more accurately say his Ireland-view) which this context would provide. The *Melodies* would thereby take their place alongside Moore's satirical and political verse (notably *Corruption and intolerance*, subtitled 'Two poems, with notes, addressed to an Englishman by an Irishman' [1808]), in addition to such works as *Lalla Rookh* (1817); they would also gain from some address upon Moore the biographer (as in his biographies of Edward Fitzgerald [1831] and Byron [1832–5]). It seems to me significant that these publications, along with his rare but consistent political interventions (as in *A letter to the Roman Catholics of Dublin* [1810]), appeared over the same period in which the *Melodies* came into being (published between 1808 and 1834 in ten volumes). The significance is that Moore can be read not as a sentimental purveyor of Irish romanticism, but as a lyric poet whose musical, biographical and indeed historical achievements were not incidental but formative in relation to that image of Ireland which the *Melodies* collectively and individually represent to the world. A collected edition, in short, would provide a much sharper perspective on Moore than we have had to date.[6]

A second perspective, likewise little explored, concerns Moore's impact on European romanticism and the impact of Europe on Moore. *Lalla Rookh* is the obvious source of Moore's influence in this regard, if only because it inspired at least one major work in the European repertory (Schumann's op. 50, *Das Paradies und die Peri*), in addition to Berlioz's debt to the Irish writer. 'The last rose of summer' is at the heart of Friedrich von Flotow's *Martha*, to say nothing of those Irish composers who directly engaged with the *Melodies* as editors or arrangers, including Michael Balfe, Charles Villiers Stanford and William Vincent Wallace. With regard to the *Melodies* as a self-standing body of work, the fact of European (especially German) interest in Celticism is strikingly important. In this respect, the connections between Beethoven and the Scottish publisher George Thomson establish an instructive point of comparison for the relationship between the publishers William and James Power, Moore and the composer John Stevenson.[7]

6 It does seem worth emphasizing that the long period over which the *Melodies* were published also marks the appearance of so many other works by Moore: neither have the *Melodies* themselves enjoyed a scholarly edition despite their being almost constantly in print since they were first published. The existence of so many partial editions of Moore's works (some within his own lifetime) makes the prospect of a complete edition all the more daunting. The Thomas Moore materials deposited in the Special Collections Department, Queen's University Belfast might well provide the impetus for producing a Complete Works, but to judge by the plenitude of individual sources and editorial techniques required, such a task would entail the cooperation of a very large editorial board, to say nothing of the substantial funding which the project would require.

7 Although I briefly discuss Beethoven's Irish folksong settings below, together with the influence of Moore on Berlioz (both of which have attracted scholarly attention), a comprehensive study of Moore's impact on European musical Romanticism, including his presence in Irish, English and German opera, is still lacking. The question of Schumann's reception of *Lalla Rookh* (via a German translation) would seem to be especially worth investigating. For a recent overview of the literature which has grown around the subject of music in German Romanticism, see Robin Elliott's introduction to Siobhan Donovan and Robin Elliott (eds), *Music and literature in German Romanticism* (Rochester, NY, 2004), pp. xi–xxix.

Beyond questions of musical influence, there arises some consideration of Moore's remarkable influence in the formation of European nationalism, especially in Poland. The cultivation of a nationalistic aesthetic in Polish literature, particularly in the verse of Adam Mickiewicz, who translated and indeed imitated Moore's verse, owes much to the precedent and influence of the Irish writer.[8]

The third and final perspective – and perhaps the most important of the three – is provided by history itself. I apologize if it seems pedantic to underline a distinction between the formative history of ideas which influenced the production of Moore's writings and the reception history which subsequently covered them in neglect or exposed them to the charge of sentimental pandering. But that distinction to one side, it is only fair to acknowledge that just as the rebellions of 1798 and 1803, together with Moore's encounter with the United Irishmen during his years in Trinity, were decisive in his formation as an artist, the reaction which his work drew from Young Ireland, as from individual partisans on either side of the political divide (Ferguson and Mangan are good instances), is expressive of a history of Irish ideas which is central to the meaning and purpose of Moore's work. To make a case for the *Melodies* in such terms is surely preferable to extolling their merits in a cultural climate long since impatient of Moore's romantic vision, to say nothing of his political tolerance and pluralism. But if nothing else, the significance of their publication deserves to be redeemed from the sentimental reputation under which they have laboured for so long.

II

If antiquarianism (the recovery of the past and of the Gaelic past in particular) was overtaken by Romanticism (the ideology of the present in relation to that past) in the first decade of the nineteenth century, then that movement was nowhere more clearly traced than in music. Thomas Moore's alignment of the ethnic repertory of music with his own poetic was the decisive episode in this process. He formulated this alignment

8 See Eoin MacWhite, 'Thomas Moore and Poland', *Proceedings of the Royal Irish Academy,* 72 (1972) sect. C, 49–62.

no more expressly than in the 'suppressed preface' to the third volume of the *Melodies*, which appeared anonymously as an appendix to the poem 'Intolerance':

> But the sceptic is scarcely to be envied who would pause for stronger proofs than we already possess of the early glories of Ireland; and were even the veracity of all these proofs surrendered, yet who would not fly to such flattering fictions from the sad degrading truths which the history of later times presents to us?
>
> The language of sorrow, however, is in general best suited to our Music, and with themes of this nature the poet may be amply supplied. There is scarcely a page of our annals that will not furnish him a subject, and while the national Muse of other countries adorns her temple proudly with trophies of the past, in Ireland her melancholy altar, like the shrine of Pity at Athens, is to be known only by the tears that are shed upon it.[9]

The admixture here of romance and history, of a grim realism in the face of Ireland's present condition by comparison with the consolations of myth and glorious reputation is, of course, a classic trope of Romanticism that is anything but unique to Moore except for its explicit degree of expression. To write thus in the aftermath of 'that time of torture and terror' (which is how Moore was to characterize his years in Trinity because of the interrogation and summary execution of United Irishmen from within the circle of his own friends) was not the pliant posture of a drawing-room romancer, but the passionate avocation of music as a conduit of political sensibility. In a letter to his collaborator on the early volumes of the *Melodies*, Sir John Stevenson, published as part of the advertisement to the first number, he palliates this insistence with a more general but no less committed observation:

> Our National Music has never been properly collected, and while the composers of the Continent have enriched their operas and

9 Cited in Deane, 'Thomas Moore', p. 1057

> sonatas with melodies borrowed from Ireland … we have left these treasures to a great degree unreclaimed and fugitive. But we are come, I hope, to a better period both of politics and music; and how much they are connected, in Ireland at least, appears too plainly in the tone of sorrow and depression which characterizes most of our early songs.[10]

This immediate identification of music with politics was to prove emblematic for Moore's collection as a whole. It is not simply that Moore's encounter with the repertory was so intimately bound up with insurgency and the mortal danger in which this placed his friends: it is also that Moore's romantic projection of sources (which were themselves entirely innocent of political meaning) was a deliberate and strategic invention that rapidly acquired symbolic and conventional status. Not only do we find that the figures of Lord Edward Fitzgerald and Robert Emmet haunt the early numbers of the *Melodies* (so that an epigraph from 'Oh blame not the Bard' appeared on an anonymously circulated print of Emmet's 'Speech from the Dock' in 1811); we can also recognize that the *Melodies* (by which I mean Moore's conjunction of new verse with pre-existing tunes) comport themselves so as to sustain this connection between music and politics, however softened the join may have appeared to be. It is idle to complain that Moore's drawing-room decorum sentimentalized his reading of history when the *Melodies* themselves demanded not the rhetoric of sedition but the currency of a well-wrought song. Sedition, albeit of a pretty muted kind, would come later, in the shape of Thomas Davis' ballads published in *The spirit of the nation*. So, too, would a rejection not only of Moore, but of any reliance upon what Davis called 'paltry, scented things from Italy'.

Beyond the drawing-room, beyond the confines of the romantic song, Moore could express more radically his detestation of that synthesis of colonial politics and religious imposition which defined for him the eponymous intolerance of life in Ireland under the penal laws:

10 See Wilfred S. Dowden (ed.), *The letters of Thomas Moore*, 2 vols (Oxford, 1964), i, pp. 116–17. The letter is dated in this edition as 'Feb. 1807.'

> ... to think that such a blooming part
> Of the world's garden, rich in nature's charms
> And fill'd with social souls and vigorous arms,
> Should be the victim of that canting crew,
> So smooth, so godly – yet so devilish too;
> Who, arm'd at once with prayer-books and with whips,
> Blood on their hands, and Scripture on their lips,
> Tyrants by creed, and torturers by text,
> Make *this* life hell, in honour of the *next*![11]

It is not my intention here to trace in any detail the publication history of the *Melodies*, but the genesis of Moore's collection deserves some comment nevertheless.[12] We know that it was William and James Power who approached Moore early in 1807 with a view to his writing words (along with other poets) for a collection of Irish airs to be set with 'symphonies and accompaniments' by John Stevenson. We know, too, that Moore had earlier been invited by George Thomson similarly to contribute to a volume which Thomson was planning, in succession to his earlier collections of Scottish and Irish airs. We know, finally, that two of Thomson's principal musical collaborators were Franz Joseph Haydn and Ludwig van Beethoven.

In 1805, Moore had written as follows to Thomson:

> Sir, I feel very much flattered by the application with which you have honored me, and the idea of being associated in any manner with Haydn is too tempting to my vanity to be easily resis-

11 From *Intolerance, a satire*, published in A.D. Godley (ed.), *The poetical works of Thomas Moore* (London, 1910), pp. 137–8, here, p. 138. On pp. 139–40 of the same edition, Godley publishes the 'Appendix' on the history of Irish music and cites Moore's observation that it was 'originally intended to form part of a Preface to the Irish Melodies; but afterwards, for some reason which I do not now recollect, was thrown aside' (p. 139).

12 For an account of the sources and publication history of the *Irish Melodies*, see Veronica Ní Chinnéide, 'The sources of Moore's melodies', *Journal of the Royal Society of Antiquaries in Ireland,* 89 (1959), 109–34 and Thérèse Tessier, *La Poésie Lyrique de Thomas Moore (1779–1852)* (Paris, 1976), passim.

> ted. At present, however, I am so strictly pledged not to divert one moment from the poems I am engaged in.[13]

As it transpires, Moore did not fulfil this commission for Thomson, but we may suppose that the prospect of doing so would have influenced his approach to the verses he wrote for William and James Power. His ready appreciation of Haydn is especially significant in this respect.

Barry Cooper remarkably informs us that 'Beethoven composed more folksong settings than any other genre, and ... more of them are Irish than any other nationality.'[14] Responding to an invitation from Thomson to contribute settings in succession to Haydn, Beethoven first contributed arrangements of Irish, Welsh and Scottish melodies to Thomson in 1809 and 1810. In 1814 and 1816 respectively, Thomson published two volumes of Irish airs with arrangements by Beethoven. In all, the composer provided seventy-one settings to the Scottish publisher. Barry Cooper makes it plain, however, that Beethoven was not setting texts, which Thomson instead added on receipt of the arrangements. This detail is a crucial one, because Moore's contemporary settings represented a collaboration in which the creation of new texts was the primary consideration, and in which the musical settings were correspondingly a secondary, if vitally important undertaking. Moore selected the airs, determined their pitch and modified their rhythmic contours to suit his newly-composed verses. He also did so to suit his own capabilities as a performer. Only then was Stevenson invited to arrange the airs and preface them with modest 'symphonies' or introductions and postludes for pianoforte. In musical terms, as I have argued elsewhere, it is the hapless Stevenson who appears gauche and intrusive in terms of this extraneous material, just as it is Stevenson's contribution which motivated so many subsequent editions of the *Melodies* throughout the nineteenth century.[15] What I should wish to identify here, however, is the

13 See Dowden (ed.), *The letters of Thomas Moore*, i, pp. 108–9.

14 Barry Cooper, 'Beethoven's folksong settings as sources of Irish folk music', in Patrick F. Devine and Harry White (eds), *The Maynooth International Musicological Conference 1995, selected proceedings: part two* (Dublin, 1996), pp. 65–81; here, p. 65.

15 See White, *The keeper's recital*, p. 48ff.

fundamental difference between Moore's settings and those of every other 'Hibernian' collection of airs, including those provided by Beethoven for George Thomson. In this connection, it is useful to turn back to Barry Cooper:

> Being from a non-Irish background, Beethoven was actually in a better position to appreciate local idiosyncrasies than someone brought up amongst them. Moreover, his instrumental introductions and postludes are closely based on material from the melodies themselves; unlike other harmonizers, including Bunting, he does not normally introduce extraneous, classical-style passages into the preludes but builds them up almost entirely from prominent motifs in the melodies … Indeed, it is only in Beethoven's settings that the rich musical potential of these Irish melodies is properly revealed, with material from them being developed in the ritornellos in extraordinarily imaginative ways. It takes a Beethoven to demonstrate what fine melodies they are.[16]

Without wishing to seem chauvinistic, or still more to criticize unduly a major Beethoven scholar, I find nevertheless that this passage betrays a major difficulty in understanding either the *Melodies* or the significance of Beethoven's folksong arrangments in terms of Irish music. Not that I would plead for one moment that Irish music deserves some kind of special exception to the general rule, but simply that Beethoven's 'appreciation' of the *Melodies* could scarcely be ranked according to his ignorance of them, no more than a German writer without English could possibly hope to comment authoritatively on the poetry of Wordsworth, still less translate him into German. The proposition that 'the rich potential of the melodies' is reserved to Beethoven's businesslike settings is also an astonishing claim, and certainly one which seems to have been made without the smallest acquaintance with how these melodies have been represented by Irish musicians in the past thirty years. Nevertheless, I

16 Cooper, 'Beethoven's folksong settings', p. 80.

dwell on these misprisions (if such they be) only to illustrate the gulf that lies between the essence of Beethoven's achievement (in which folksong arrangements are a peripheral activity) and the true significance of Moore's address upon this repertory. It is not simply that the Beethoven commissions are at best a footnote to the composer's contribution to the European repertory (so that the mere number of seventy-one folksong settings exceeding any other genre in Beethoven's output becomes in turn a meaningless statistic), but also that Moore's arrangements went to the heart of Irish music as an intelligencer of verbal feeling. And not just verbal feeling in general, but verbal feeling in English and specifically in regard to the enduring tropes of loss and dispossession.

The traditional point of comparison between Moore and his predecessors, has of course been hitherto largely limited to Bunting. Some years ago, I tried to define the difference between the two men by suggesting that Moore's success as the 'Bard of Erin' transformed Bunting's *Ancient Irish music* from antiquarian collection to Romantic source. It was not a permanent transformation, and in antiquarian terms, it remained an indefensible one. But we cannot disclaim the fact that Moore's poetic was advanced and developed by his fantastically successful recourse to the melodies which Bunting had so painstakingly (if intermittently) struggled to preserve. At the last, the difference between them was that Moore became interested in the past as an image of the present, whereas Bunting was committed to the past for its own sake. It would not be too much to add that certain strains of cultural history and folklore respectively maintain this difference even today. What is clear is the continuity of that relationship – musically, culturally, aesthetically and politically – between Moore and his antiquarian predecessor, a relationship that could not be usurped by the routine commissions which a Scottish publisher sought from a German composer living in Vienna and utterly aloof from the orbit of Irish sensibilities.[17]

Moreover, in ways that neither Thomson nor his composers could conceive, the response to early numbers of the *Melodies* replicated the

17 For a more detailed consideration of the cultural differences between Bunting and Moore, see White, *The keeper's recital*, p. 50ff.

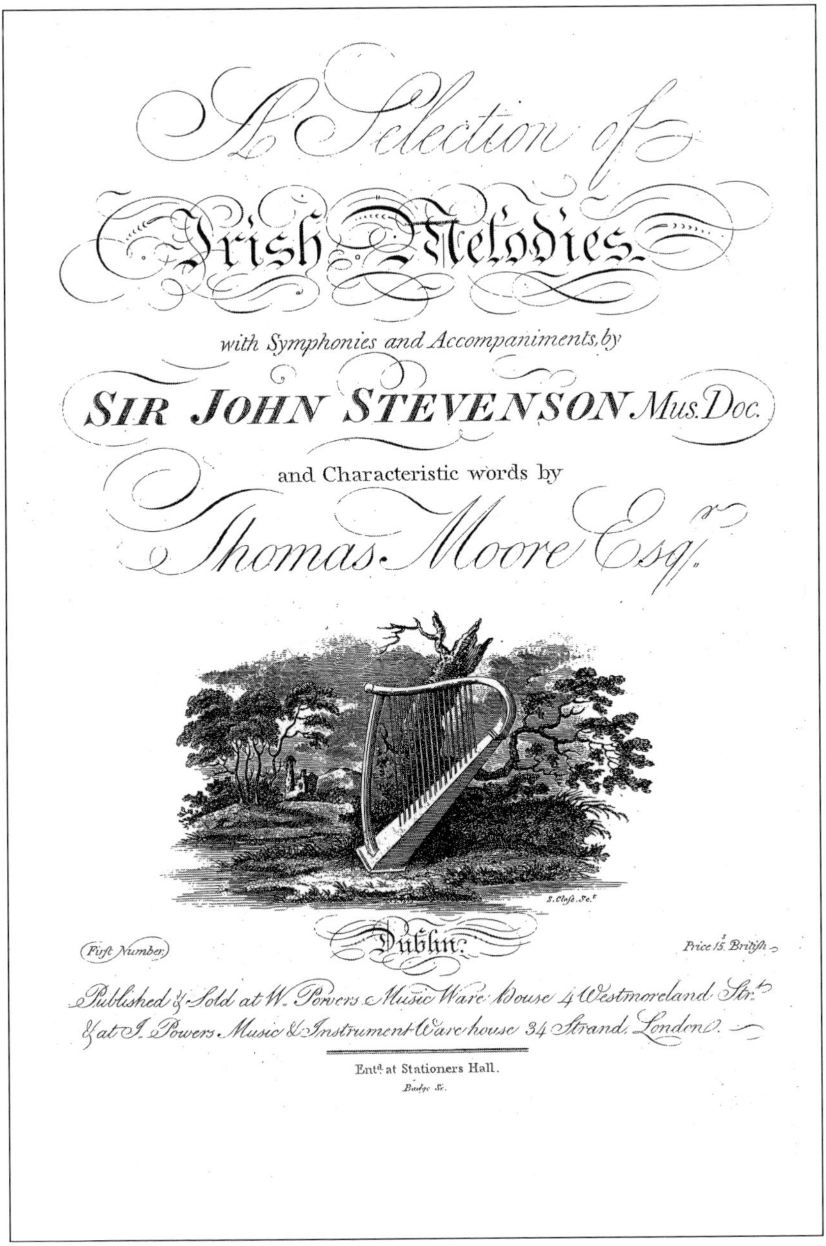

2 Title page of the first number of the *Irish melodies* (Dublin & London, 1808)

divide that even then characterized the Protestant Interest in Ireland and its unruly, papist counterpart. The *Anti-Jacobin Review* was predictably severe:

> Several of them [the *Melodies*] were composed with a view to their becoming popular in a very disordered state of society ... They are the melancholy ravings of the disappointed rebel, or his ill-educated offspring.[18]

Much in this vein was to follow: if the *Melodies* would never quite shake off the aura of the drawing-room, neither would they lose their symbolic force as a cumulative petition for self-regulation. Through the course of their publication, this association gathered pace. Moore's biographer, Hoover Jordan, reports that as early as October 1809, a writer in the *Leinster Journal* referred to Moore as 'that poet of the heart, who has done more by his poetic effusions than all the political writers who[m] Ireland has seen for a century'. By 1841, the *Dublin Review* was prepared to admit of Moore's political influence in surprisingly candid terms:

> We have somewhere seen it asserted that the influence (which all must admit) of the Irish *Melodies* in advancing the great cause of Catholic emancipation was exerted in the higher circles of English society, where the language of them and the sweet music in which it was wedded, excited a sympathy never before felt for the suffering country.[19]

The notion of Moore as an advocate of Catholic emancipation through the *Melodies* would in turn be eclipsed by their very popularity and by the overwhelming presence of O'Connell's more sectarian agitations: nevertheless, this trope of emancipation directly nourished that projection of Moore as an apologist for nationalist self-determination which

18 *Anti-Jacobin Review*, 58 (1820), p. 314, cited in Hoover H. Jordan, *Bolt upright: the life of Thomas Moore* (Salzburg, 1975), p. 158.

19 Cited in Jordan, *Bolt upright*, p. 144

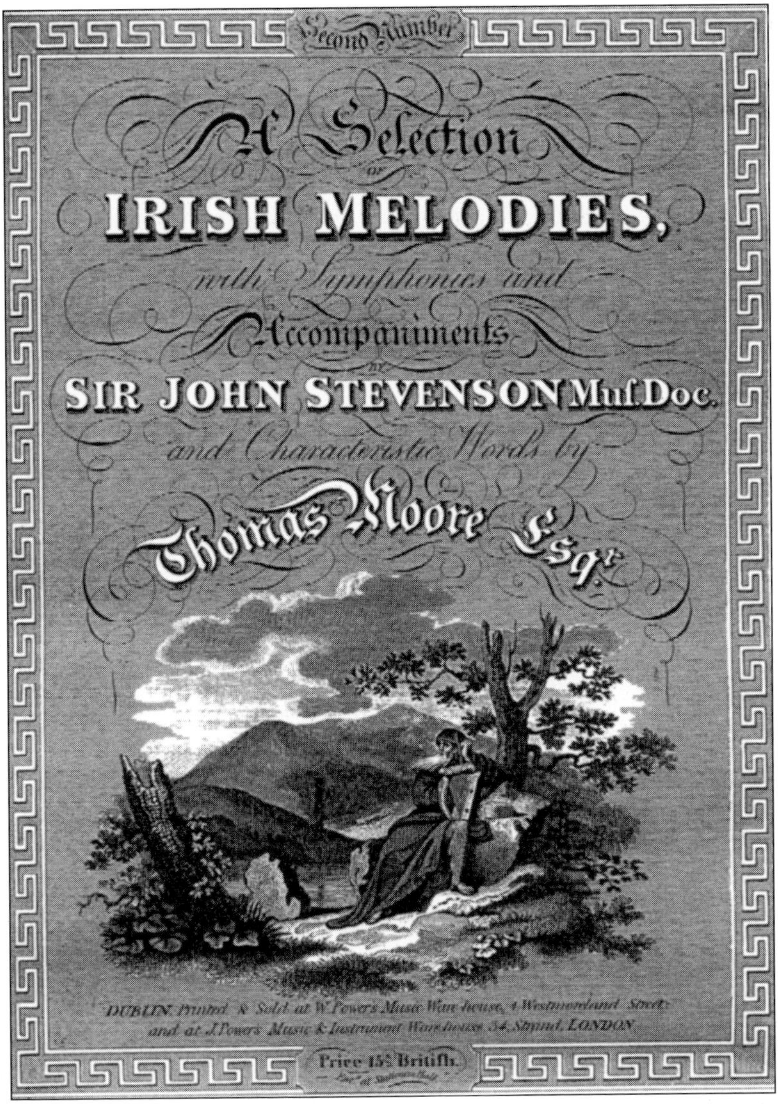

3 Title page of the second number of the *Irish melodies*
(Dublin & London, 1808)

Young Ireland would in turn come to adopt. Although the currency of this idea stimulated opposition (from Bunting no less than from Samuel Ferguson and James Clarence Mangan) we find that Gavan Duffy, a principal architect of Young Ireland, was prepared in 1842 to acknowledge (in defiance of his mentor, Thomas Davis), precisely that reading of the *Melodies* which Moore himself appeared to sanction in his prefatory remarks to the very first number. Duffy's apologia is worth quoting at some length:

> 'And can this', thought we, taking up the *Melodies* – 'can this be forgotten?'
>
> No; these songs are no longer as they were, popular only in the drawing-rooms of Europe and America; they are gradually becoming known to the middle classes in Ireland, and the Irish translation bids fair to reach the mind of our peasantry. It may be fault or excellence in them; but Moore's songs bear translation. They not only have appeared in every European language, but they supplied the Poles with their most popular revolutionary songs during the last war – the highest honour ever shown to a lyrist.[20]

Duffy refers here to the Irish translation by John McHale, archbishop of Tuam, of the texts of the *Melodies* in 1842, and also to the wider dissemination of Moore's verse not only across class divisions in Ireland but across Europe. More explicitly still, his acknowledgment of Moore's influence on the formation of Polish nationalism (through the writings and translations of Adam Mickiewicz) in the revolution of 1838, signals a vital distinction between verbal feeling and the music that by now was all but synonymous with the definitive expression of that feeling.

There can be little doubt that this synonymous condition of feeling, whereby not only Moore's *Melodies* but Irish music in general became both expressive and symbolic of dispossession and loss, survived the Great Famine. This endurance, however, did not prevent the

20 Charles Gavan Duffy, 'Thomas Moore' (1842), quoted in Deane (ed.), *The Field Day anthology of Irish writing*, i, p. 1251.

Melodies, now largely divested of Moore's wider achievements as a writer, from falling headlong into the cloying divide that lay between the integrity of Gaelic revivalism and the aspirations to develop an art music at once expressive of Ireland and worthy of Europe. Between these extremes lay the balladry of popular agitation, the sporadic presence of art music (as against its otherwise putative existence) and the representation of Moore as a pro-British sentimentalist or wailer after a lost cause.

III

The appearance of innumerable popular editions of the *Melodies* throughout the nineteenth century hastened Moore's swift declension from national poet to the embodiment of a popular culture from which both the Gaelic League and the Celtic Revival were swift to distinguish themselves. I have argued at length elsewhere that Moore's uneasy posthumous relationship with the Celtic Revival and modern Irish writers in general entailed a repudiation which was as widespread as his earlier acclaim had been.[21] When he resurfaces amidst the flotsam and jetsam of consciousness in *Ulysses* and *Finnegans Wake*, he is consigned to the detritus of Irish cultural discourse, as irrelevant to the modernism and myth making of Joyce's recrudescence as he is to the brilliant recovery of Gaelic mythology in Yeats. Located perilously between the claims of music (which could only be countenanced by the Revival as a symbol of the literary imagination itself) and those of poetry and drama, Moore's *Melodies* ('tummy moores maladies', in Joyce's half-affectionate, half-sarcastic formulation) lapsed into disrepute. But if writers establish their own forbears, it seems only right to add – as Seamus Deane has done – that it is precisely Moore's musico-poetic conjunction of Irish music and letters that restores to the *Melodies* their seminal role in the transformation of Irish literature written in English. Divested of their sentimental reputation (to say nothing of Moore's own projection as an eager socialite in love with the upper reaches of English life), the songs might well have

21 See White, *The keeper's recital*, p. 151ff.

been regarded not as a cul-de-sac, but as a vital precedent for the revival itself. But by the end of the nineteenth century, the polarized condition of music in Ireland – either as a symbol of Gaelic culture or as a shibboleth of colonial imposition – meant that the *Melodies* instead entailed a reception history of some contempt amidst the literati of the new Ireland.

Their afterlife in the attempts to establish a continuous Irish art music was, however, otherwise. Entirely fresh editions appeared in 1859 (with new 'symphonies and accompaniments' by Balfe – another victim of Joyce's demotic understanding of musical culture) and in 1894, by Charles Villiers Stanford.

Stanford's encounter with Moore was decisive. Hugely prolific as a composer throughout his long career (it spanned from the late 1860s until shortly before his death in 1924), Stanford's complex relationship with Ireland is a matter which can scarcely be addressed here, except to note that his absorption of indigenous Irish music was more thoroughgoing than any composer before or after him, with the sovereign exception of Seán Ó Riada. At first glance, this may seem a strange assertion, but it simply happens to be true. Stanford's monumental edition of the Petrie collection of Irish music (1902–5), his 1887 publication *Songs of old Ireland* and his edition and arrangement of the *Melodies* (entitled, significantly, *The Irish melodies of Thomas Moore, the original airs restored*) in 1894 can be regarded, as his recent biographer Jeremy Dibble puts it, as 'an immensely important formative event in the development of Stanford's musical personality'.[22] To recover the terms advanced by Seamus Deane quoted at the outset of this essay, the 'metaphor of the translation of Catholic into Protestant, of native and antique authenticity into modern and equally native civility' which Deane applies to Moore himself can equally well be applied to Stanford's attempts to come to grips with the ethnic repertory, through the medium and precedent of Moore's own musical 'translations'.[23]

The problem was that the quotation of the melodies (often in full)

22 Jeremy Dibble, *Charles Villiers Stanford: man and musician* (Oxford, 2002), p. 129.

23 For a discussion of Stanford's recourse to Irish music in general and Moore in particular, see Dibble, *Charles Villiers Stanford*, p. 128ff.

across the entire range of Stanford's output represented a failure on two counts. First, this compositional habit repeatedly manifested an inability to integrate the intonations of Irish music with the organic structures of symphonic discourse (with the result that Irish melodies from Moore and from other collections tend to stand out with a kind of exotic self-consciousness from the surrounding Germanic context in Stanford's instrumental music). Secondly, Stanford, so utterly unlike Moore, was an Anglo-Irish unionist whose enchantment with the European aesthetic in music fortified the association which many in Ireland would draw between art music and the colonial presence. Nor could he supervene the (by then) depleted condition of Moore's reputation, to the extent that any kind of Irish music, and notably the 'drawing-room' finesse of Moore's arrangements, spoke to an Ireland which had long ceased to be of interest to Stanford's contemporaries in literature and the visual arts. Although his influential incorporation of Irish folksong strongly characterized the perception of Irish art music, so that a debate (as yet unresolved) as to the relevance of folksong quotation would dominate discussions of this matter well into the twentieth century, the 'transformation' of which Deane speaks was, in his case, an unsuccessful one.[24]

Many composers were to follow Stanford's powerful lead in quoting Moore (or similar melodies) as the vital hallmark of authenticity in Irish music, and all encountered the same dissatisfying results. What I have elsewhere described as the 'dutiful presence of the Irish air' proved to be an obstacle which Irish art music would not overcome for decades after Stanford's death, with the notable exception of Arnold Bax, whose music bypassed the legacy of Moore and uniquely depended instead on the Celtic Revival and on Yeats himself. Thus the dismal reading of posterity: just as he had been consigned to the limbo of sentimental popularity by the politics and culture of the Celtic Revival, so too Moore's afterlife in Irish art music proved to be a hindrance rather than a help.

24 For further discussion of Stanford's failures in regard to the use of Irish folk music (including the *Melodies*), see Harry White, 'Art music and the question of ethnicity: the Slavic dimension of Czech music from an Irish point of view', *International Review of the Aesthetics and Sociology of Music*, 35:1 (2004), 29–46.

By the time Seán Ó Riada had rejected European music as being 'not Irish', Moore's address upon the ethnic repertory was utterly marginalized and – it must be admitted – most brilliantly usurped by the virtuosity and ensemble panache not only of Ó Riada but of Ceoltóiri Cualann, the Chieftains and the polished exponents of Irish traditional music in the late twentieth century.

It is impossible to speculate as to what might have happened to the reception of Moore in a country more engaged with art music and less polarized by the music question than was Ireland in the nineteenth century. But Moore's reception in Europe gives us some indication of his inherent musical and literary status, independent of those ideas which dominated his reputation at home. I have already mentioned in passing Schumann's setting of *Lalla Rookh*, but French music offers a more pertinent instance in respect of the *Melodies* themselves. Julian Rushton tells us that in 1829, Hector Berlioz, whose opus 1 bore an epigraph from Moore, published nine songs under the title 'Irlande': Rushton shows, moreover, that the French composer was explicitly influenced by Moore's word setting, although he composed new melodies – in express imitation of Moore's style – to French translations of Moore's verse. Rushton further reminds us that Moore's reputation in Paris by the early 1820s was such that the French term *mélodie* (as an equivalent to the German *Lied*) derived directly from Moore, and that Moore himself must be counted 'at this time a literary and musical influence on the level of Goethe and Shakespeare'.[25] It is not enough, it would seem, to attribute this influence simply to Berlioz's obsession with Harriet Smithson and all things Irish, but to the enormous circulation of Moore's collections and indeed to the impact of Moore's own performances of the *Melodies* in the 1820s. Finally, Rushton notes that 'the appeal of Moore's Irish *Melodies* is also attributable to their intrinsic qualities and particularly, as was claimed in an early review, an imprint of melancholy even on the cheerful songs; the same review mingled sentiment and politics by claiming that the Irish *Melodies* give a complete picture of the customs, his-

25 See Julian Rushton, 'Berlioz and *Irlande*: from romance to *Mélodie*' in Devine and White (eds), *The Maynooth International Musicological Conference 1995*, pp. 224–40.

tory and suffering of the Irish people.'[26] In such an immediate reception history, it is possible to see how Moore's romantic nationalism might be integrated into the wider discourse of European art music.

'Every home should have its copy of Moore and Davis', advised Douglas Hyde in his 1894 lecture, 'The necessity for de-Anglicizing Ireland', a prescription which makes the *Irish melodies* sound like a harmless but unmistakable precursor to *Mein kampf.* The fact remains, however, that the ideas of Ireland and of Irish music which the *Melodies* released into the cultural stratosphere were so powerful – specifically in regard to the conventional understanding of Irish music as the primary intelligencer of dispossession, loss and identity – that once identified, they overtook Moore's own tolerant pluralism, to say nothing of the promise which the songs themselves held in purely musical terms. At the last, we must distinguish clearly between Moore's achievement in creating the *Melodies* and the astonishing influence – not all of it positive – which they exerted on the cultural regeneration of Irish music and literature in the later nineteenth and twentieth centuries. In this respect, our watchword should probably be, 'Oh blame not the Bard'.

26 Rushton, 'Berlioz and *Irlande*' (p. 226) quotes from Allen B. Thomas' *Thomas Moore en France* (1911).

Cabaret, sex and independence

Publishing in the early Free State

NICHOLAS ALLEN

IRELAND'S FIRST DECADE of political independence is typically perceived to have been a barren period of depression and disillusion, the toll of revolutionary struggle, civil war and state foundation weighing upon government and citizen alike. Writers, we imagine, were marginal figures, having surrendered to censorship or exile. But there is evidence, fragmentary but suggestive, to support a view of Dublin in the 1920s as a city where artists, actors, musicians and politicians mingled in the margins of a miniature Berlin, experimenting in radical forms, burlesque and low comedy, pamphlets and periodicals, the erotic and irreverent – all expressions of a society newly mutable in a Europe undergoing major transformation following the First World War. With its revolutionary glamour, Ireland was a centre of European re-imagination. Shortly before his death, Marcel Proust rebuked a visitor who asked him about his work, saying that Terence MacSwiney was a subject more fit for conversation. 'Don't speak to me ... about *Du côté des Guermantes*; but about the Lord Mayor of Cork,' he replied, 'that will be very interesting'.[1]

Irish intellectual life of the 1920s was dominated by William Butler

1 Quoted by J.M. Hone in 'A new book of snobs', *Dublin Magazine* 1:2 (Sept. 1923), 221. I am grateful to Brendan Barrington for his comments regarding this essay.

Yeats – so much so that a group of younger artists, centred in the first instance on Liam O'Flaherty, chose to assert their own interests by the foundation of a Radicals' Club. Through cabaret and polemic, O'Flaherty and his contemporaries prepared audiences for the reception of an avant-garde that, by the 1930s, included Samuel Beckett, James Joyce, Thomas MacGreevy and Blanaid Salkeld. The founding members of the Radicals' Club were O'Flaherty, the painter Cecil Salkeld, Austin Clarke, Gabriel Fallon, F.R. Higgins, and Brinsley MacNamara. The club's first meeting was planned for Sunday, 2 August 1925 at 6.30p.m., with O'Flaherty expecting fifteen people each willing to pay £1 subscription to be present.[2] Sean O'Casey was invited to join. O'Casey refused, later writing he

> would argue with Yeats, oppose him manfully, but personally, on any question of religion, politics, and even literature, if he happened to differ from the poet; but he wouldn't join any clique to do it, because he thought this opposition was born of envy of the great fame the poet enjoyed as the leading man of Irish Letters.[3]

O'Flaherty was unperturbed by O'Casey's demurral, emboldened as he was by the publication of his new, and popular, novel, *The informer*. O'Flaherty was optimistic of the club's potential, but keen 'to keep out Yeats, Senator Gogarty, etc., who want to have their finger in every pie and make a mess of it'.[4] He wrote to the English critic Edward Garnett in a state of high excitement soon after the club was established, having gathered 'all the intellectuals into one club … that in itself should cause such an uproar that the government will have to call out the troops to save the last of the ladies' hair and of the men's whiskey'.[5] Austin Clarke's more cynical eye pierced this flattering self-projection. Writing to F.R.

2 F.R. Higgins correspondence, National Library of Ireland [hereafter NLI] MS 10864/1/9.

3 Sean O'Casey, *Inishfallen, fare thee well* (1942; repr. London, 1992), pp. 153–4.

4 Liam O'Flaherty to Edward Garnett, 6 Sept. 1925 in A.A. Kelly (ed.), *The letters of Liam O'Flaherty* (Dublin, 1996), p. 130.

5 Liam O'Flaherty to Edward Garnett, 5 Oct. 1925 in Kelly (ed.), *The letters of Liam O'Flaherty*, p. 134.

Higgins of a visit to O'Flaherty, Clarke found O'Flaherty 'quite a nice fellow as you said: but I was bored: all time talk of drink and horses: neither of which materialized, only tea'.[6]

High living was reserved for club meetings on Sunday evenings. We might indulge a natural scepticism if accounts of this group came from O'Flaherty himself – a writer whose mendacity in pursuit of career and establishment throughout the 1920s is shocking. Records of their activities are sparse; we do not know most of their transactions. But, fortunately, we have the testimony of one of the regular members of the Radicals' Club, Rosamund Jacob, a Waterford-born Quaker whose parents were both nationalist and atheist. Jacob's records are intriguing, and offer us a glimpse of the radicals' version of cabaret, that underground entertainment first experimented with in Montmartre's *Le chat noir* in the 1880s and developed, more pressingly still, in post-war Berlin as an art form representative of the upset energies of a society, like Ireland's, in major transition. Jacob, a novelist and historian, had been herself a member of the Gaelic League since 1906 and active in Sinn Féin from 1916; she was also long-time secretary to the Women's International League, a member of the committee for the Women's Social Progressive Movement, and a delegate, in 1932, to the Friends of Soviet Russia. She was further a feminist, anti-Treaty republican and animal rights activist. Her account of a Radicals' Club performance of March 1926, held in the immediate aftermath of the riots that attended Sean O'Casey's *The plough and the stars*, gives us some idea of the iconoclasm to which O'Flaherty aspired. Jacob remembered she

> Went looking for a club meeting … & found a Radical Club rehearsal for the cabaret going on – O'Lochlainn & Fay Sargent & later Paul Farrell doing a rehearsal of part of Macbeth, with gorgeous low dialogue interspersed, O'Lochlainn coaching the others – such magnificent lurid language I seldom heard. & the shock when they came to a bloody or a naked in the play … Finally the audience were to rise & mob them in protest & then

6 Austin Clarke to F.R. Higgins, undated, in F.R. Higgins correspondence, NLI MS 10864/1/2.

> O'Lochlainn was to imitate Yeats' speech. O'Lochlainn was adorable that night, with his severity & hells & bloodys & his granite seriousness.[7]

The 'O'Lochlainn' of the above is Colm Ó Lochlainn, proprietor of the Three Candles Press and editor of the *Irish Book Lover*, the major Irish record of bibliography and books over fifty years, and more, of publication. Like Jacob, Ó Lochlainn was a republican, and his cabaret performance is some register of shifting intellectual sympathies in the year of the foundation of Fianna Fáil. Since the foundation of the State, many senior Revival writers had sided with the new Treaty establishment; George Russell in the *Irish Statesman*, William Butler Yeats in the Senate. The Radicals' cabaret was a sign, like Fianna Fáil, of a resurgent republicanism. After taking an evening's dancing at the Republican Club on Parnell Square the following night, Jacob returned to the full performance of the cabaret two days later.

> Went early to Hardwicke Hall to get the cabaret tea, no tea was there for hours, only Seamus McCall or Bowe carpentering the stage ... I wish J Brennan was not the official humourist – her song about the gossips of Dublin wasn't much, though the singers did it well ... There was one [Irish dance], where I had Mitchell Cogley, all the rest was jazz. The Macbeth play was good if one hadn't seen the rehearsal which was far better. We rushed the stage & fought, & O'Lochlainn made his Yeats speech very well.[8]

The mock riot refers to an Irish culture of staged controversy that takes its pedigree from the plays of John Millington Synge. There is every reason to believe that the disturbances surrounding the first production of *The playboy of the western world* in 1907 were consciously inspired, if not instigated, by Synge himself, just as the Radicals' Club prepared their own riot. As the *Irish Independent* reported of the *Playboy*, protesters in

7 Rosamund Jacob diary, 4 March 1926, 51. NLI MS 32582/47.
8 Ibid., pp. 58–9.

> the pit and gallery signalized their success by triumphant yells, shouts, tramping of feet, and the belabouring of seats and walls with sturdy sticks, mingled with the singing of 'The West's Awake', 'A Nation Once Again', and such like well-known patriotic compositions, in which nearly all the audience joined. At ten o'clock, the curtain was raised on the second act. Once more the performance was merely pantomimic owing to the hurricane of hissing with which it was greeted.[9]

Pantomimic is the key word here, as Synge himself realized. Interviewed immediately after the furore, the playwright explained the female characters' attraction to Christy Mahon as a ruse 'to bring out the humour of the situation'.[10] The play was 'a comedy, an extravaganza, made to amuse'.[11] This burlesque tradition of Irish drama survived Synge, notably in the drama of Gerald MacNamara, the northern playwright whose *Suzanne and the sovereigns* unites William of Orange and James II under the aegis of the Christmas fairy.[12]

The absurd, in its sense as disharmony, was a cultural form manifest through the period from Synge to Jacob and includes that most unreasonable (as most idealistic) gesture of all, the 1916 proclamation of the Irish Republic by Patrick Pearse. To the newspapers that reported the rebellion, there was disbelief in Pearse's assertion of self-determination (and disbelief was general; Frank Robbins, a volunteer in the Irish Citizens' Army, remembered in *The starry plough* how a British army officer was only persuaded of a revolution in progress when Robbins' detail frightened the officer's horse with a bayonet prick to the backside). There had been equal disbelief, expressed in anger, at Synge's declaration of Irish women's love for a parricide. Pearse, for his part, chose

9 *Irish Independent*, 28 Jan. 1907, quoted in James Kilroy, *The 'Playboy' riots* (Dublin, 1971), pp. 16–17.

10 *Dublin Evening Mail*, 29 Jan. 1907, quoted in Kilroy, *The 'Playboy' riots*, pp. 24–5.

11 Ibid.

12 For further discussion of MacNamara's work see Mary Burgess, 'Belfast carnivalesque: the satires of Gerald MacNamara', in Nicholas Allen & Aaron Kelly (eds), *The cities of Belfast* (Dublin, 2003), pp. 74–87.

THE KLAXON

an Irish International Quarterly, published in Dublin, concerned with the activities of all Nations in matters of Art, Music and Literature, appears each season.

The Editors invite contributions in verse and prose from persons in sympathy with the ideals of this journal. International correspondence on modern art movements welcomed. Payment for contributions depends upon the success of the project.

Subscriptions for One Year, 4/9 (foreign money orders accepted), to be addressed to the Circulation Manager, "The Klaxon," 179 Great Brunswick St., Dublin.

NEXT SEASON'S KLAXON includes: Article on Modern Art, by Paul Henry; Teatro Grottesco, by Walter Starkie.

CONFESSIONAL

WE are the offspring of a gin and vermouth in a local public-house. We swore that we were young and could assert our youth with all its follies. We railed against the psychopedantic parlours of our elders and their old maidenly consorts, hoping the while with an excess of Picabia and banter, a whiff of Dadaist Europe to kick Ireland into artistic wakefulness.

We produced our seven articles of faith: announcing primarily our belief in ourselves and a catholic æstheticism that would include the xylophone as well as the spinet. Picasso and Ingres, Shakespeare and Aldous Huxley, Beethoven and the Organ Grinder, Chaplin and Chaliapin were mingled in our incredible credo. We put the psychologist on his knees before the gymnast and punched our fellows into believing in the divine right of artists. We put the four corners of the world round Ireland and clacked our heels together with merriment at the resultant *macédoine de fruits*.

I

1 First page of *The Klaxon* (Winter 1923–4)

his stage well: the General Post Office, symbol of a communication system that would carry his message, as it did, to all parts of the Empire. In expressing his beliefs on the public stage, in anticipation of changes not yet come, Pearse is the link between Synge and the radical generation of the 1920s (and it is worth recalling that O'Flaherty, leading light of the iconoclasts, was himself involved in a failed communist putsch in 1922).

This culture of manifesto extravagance fed periodicals, especially that key text of Irish modernism, the *Klaxon*. The genesis of the *Klaxon* was its founding editor Con Leventhal's ejection from the *Dublin Magazine* because printers excised his essay 'The *Ulysses* of Mr James Joyce' on grounds of propriety. Billed as 'an Irish International Quarterly, published in Dublin', 'concerned with the activities of all Nations in matters of Art, Music and Literature', the *Klaxon* – which in the event appeared just once, in late 1923 – invited 'contributions in verse and prose from persons in sympathy with the ideals of this journal. International correspondence on modern art movements welcomed'.[13] Naturally, 'Payment for contribution depends upon the success of the movement'.[14] The *Klaxon* contained ten written articles, as well as illustration. The first contribution was by Leventhal, writing under the name L.K. Emery, and entitled 'Confessional'. 'We are', he wrote,

> the offspring of a gin and vermouth in a local public-house. We swore that we were young and could assert our youth with all its follies. We railed against the psychopedantic parlours of our elders and their old maidenly consorts, hoping the while with an excess of Picabia and banter, a whiff of Dadaist Europe to kick Ireland into artistic wakefulness.[15]

Like the liturgy which this editorial mimics, the *Klaxon* had

> produced our seven articles of faith: announcing primarily our belief in ourselves and a catholic aestheticism that would include

13 'LKE', 'Confessional', *The Klaxon* (Winter 1923–4), 1.

14 Ibid.

15 Ibid.

> the xylophone as well as the spinet. Picasso and Ingres, Shakespeare and Aldous Huxley, Beethoven and the Organ Grinder, Chaplin and Chaliapin were mingled in our incredible credo. We put the psychologist on his knees before the gymnast and punched our fellows into believing in the divine right of artists. We put the four corners of the world round Ireland and clacked our heels together with merriment at the resultant *macédoine des fruits.*[16]

Leventhal's 'The *Ulysses* of Mr James Joyce' is the outstanding contribution to the *Klaxon*. Leventhal defends *Ulysses* against the charge of pornography on the grounds that Joyce explores sexuality in due relation to all human function. Leventhal's comments had a broader charge for the Irish writer: that the artist need not work in a national tradition for authenticity, and that the realist mode could include a sense of the magical, as the events of one day can include 'a world of learning, scientific, metaphysic, æsthetic, of life real and imagined, of force comic and tragic, of description, of lyricism, patter and rhetoric, heroics and eloquence'.[17] By this sense of mutable time, Leventhal echoes Einstein, to whom he refers, and Bergson, who distinguished between the durational time of the clock and the experiential time of the individual. Leventhal considered Joyce to be

> the product of his age, or perhaps, as with all genius, a little ahead of it. In him we find collected all the strivings of the modern world. That which stands out most is the kinship between him and modern painters. This year's pictures at the Salon d'Automne have precisely the same effect as *Ulysses* on the conventional mind. It calls the true ugly, because truth comes in the shape of a squatting lady with an abundance of fat.[18]

The two great successes of the 1923 Paris Salon d'Automne were the

16 Ibid.

17 Laurence K. Emery, 'The *Ulysses* of Mr James Joyce', *Klaxon* (Winter 1923–4), 4.

18 Ibid., p. 16.

Russian painter Pavel Tchelitchew and the Spanish Pedro Pruna, both of whom were influenced deeply by Picasso's interest in Dada. Dada originated in Zurich; one of its founders, Hans Arp, could 'affirm that Tristan Tzara discovered the word Dada on the 8th of February 1916 at 6 o'clock in the evening ... This took place at the Café de la Terrasse in Zurich and I had a roll of bread up my left nostril'.[19] Whatever Arp's culinary tastes we can easily see that Dada meant to shock. A prototype of surrealism, Dada was an anarchic movement that combined the irrational with polemic. As the artist Francis Picabia declaimed in his *Manifeste cannibale dada*, 'Hiss, shout, kick my teeth in, so what? I shall still tell you that you are half-wits. In three months my friends and I will be selling you our pictures for a few francs'.[20] Synge could not have put it better. Dadaists introduced scientific and geometric elements to their painting, and journals like *291*, *Der Ventilator* and *Wrong-Wrong* experimented with type-face, lettering and fragmented design, such as we find in the *Klaxon*. Leventhal made this connection explicit in his consideration of Joyce, noting 'The form of burlesque which runs through the whole of *Ulysses* is a point of contact with some Frenchmen known as Dadaists'.[21] After comparing the Cyclops section of the book, as it is now known, to a Max Jacob poem, Leventhal observed:

> Mr Joyce has a little more in common with the Dadaists: his inventions of onomatopoeic words and the mixing of science with literature. Joyce and the Dadaist arrived independently, of course, at the similarities. They merely prove how true a son of his age is the author of *Ulysses*. Laurence Sterne I have always considered the first Dadaist. He might almost be considered the christener before its birth of the movement ... Tristram happens himself at certain intervals and changes of the moon to be both

19 Quoted in Nicholas Allen, *George Russell (Æ) and the new Ireland, 1905–30* (Dublin, 2003), p. 170. This book contains full discussion of periodicals and politics in the period of Irish revolution, including the *Klaxon* and *To-morrow*.

20 Francis Picabia, *Manifeste cannibale dada*, read at the Dada soirée at the Théâtre de la Maison de l'Oeuvre, Paris, 27 Mar. 1920.

21 Emery, 'The *Ulysses* of Mr James Joyce', 17–18.

> fiddler and painter according as the fly stings. Sterne is an eighteenth-century Joyce. They both go to antiquity for quips and quibbles, and have the same taste for salaciousness and the use of names for comic effect.[22]

To Leventhal, Joyce is like Picasso, 'the early follower of tradition breaking into new modes and expressing life from a new angle with a changed vision'.[23] Admiration for Picasso is a central theme of the sole issue of the *Klaxon*. In an essay entitled 'Picasso, Mainie Jellett and Dublin criticism', Thomas MacGreevy defended Jellett as the Irish painter whose critics, antagonistic to her cubism, were 'as hopelessly at sea in front of her work as her benighted predecessors were about Picasso and Othon Frieze and Matisse in 1912'.[24] Such conservatism was evidence of a deeper malaise in the institutions of Dublin art and patronage. Echoing the sentiment of George Bernard Shaw's private comments of 1908 that 'the Dublin National Gallery' was 'one of the best in Europe of its size, and therefore less populous than Connemara',[25] MacGreevy observed:

> The National Gallery has never been a favourite haunt of Dublin people, except for an odd pair of lovers who discover what a delightfully lonely retreat it is. The Municipal Gallery has from its earliest days been popular with students as well as with the general public. For we are all – except our newspaper critics – interested in pictures by Irish artists and French artists. We are not interested in dead Dutch boors and dead English gentlemen. They only tell us about boorishness and gentility. They tell us nothing about art.[26]

Polemic in little magazines is usually forgotten beyond the first months of production. But the *Klaxon* did have one immediate impact

22 Ibid., 18. 23 Ibid.

24 Thomas MacGreevy, 'Picasso, Mamie Jellett and Dublin criticism', *Klaxon* (Winter 1923–4), 23.

25 Quoted in Allen, *George Russell*, p. 12.

26 MacGreevy, 'Picasso, Mamie Jellett and Dublin criticism', 27.

on Irish literature, the publication of *To-morrow*, which survived for two issues in August and September 1924. *To-morrow* was consciously different from its Dadaist predecessor in format, with three columns of newspaper type a gesture towards an objectivity the *Klaxon* denied. *To-morrow* contained work by Joseph Campbell, Iseult Gonne, Liam O'Flaherty, Lennox Robinson and William Butler Yeats in a style of mock reportage. Edited by Francis Stuart and Cecil Salkeld, *To-morrow* probably took its title from a play of the same name written by Henry O'Hanlon and performed by the Irish Theatre, Hardwicke Street – of which Cecil Salkeld's mother Blanaid was a member – for six nights from 18 December 1916. This work, set in a morgue, was described in the journal *New Ireland* as 'a type of play we rarely see in Ireland, but which is common enough on the Continent', a 'gruesome', if 'fascinating' meditation on life and death.[27] Such a description equally fits the journal. Yeats, willing to harness the energy of such ambition in service of his own agenda, took the opportunity of *To-morrow*'s first appearance to publish his own manifesto under the names of the editors. 'To All Artists and Writers' preaches revolution in a medieval language of eternity that is radical only in the context of Ireland's own religious conservatism, Yeats' advocacy of the soul an attempt to foreground individual freedom of perception.

> We condemn the art and literature of modern Europe. No man can create, as did Shakespeare, Homer, Sophocles, who does not believe, with all his blood and nerve, that man's soul is immortal, for the evidence lies plain to all men that where that belief has declined, men have turned from creation to photography. We condemn, though not without sympathy, those who would buy escape from banal mechanism through technical investigation and experiment. We proclaim that these bring no escape, for new form comes from new subject matter, and new subject matter must flow from the human soul restored to all its courage, to all its audacity. We dismiss all demagogues and call back the

27 Quoted in Allen, *George Russell*, p. 171.

> soul to its ancient sovereignty, and declare that it can do whatever it please, being made, as antiquity affirmed, from the imperishable substance of the stars.[28]

Yeats' vision of modernity as an endless recycling of earlier modes informs the impulse behind the most shocking contribution to *To-morrow*, Lennox Robinson's 'The Madonna of Slieve Dun', a harrowing tale of rural rape transformed by the victim's self-delusion to a sense of new annunciation. The story caused a furore on the Carnegie Library Committee, of which Robinson was secretary, forcing the resignation of Father Thomas Finlay and John Henry Bernard, the provost of Trinity College Dublin. The story is set in a Synge-like parish, beyond the rule of written law, an *aporia* where nameless misdeeds echo through community consciousness.

> Father Kearney was too old and feeble to keep a watch on the lads … every year they grew wilder and more lawless. They had to have an extra policeman at the barracks, and there was never a Petty Sessions that there weren't men had up before the magistrates for assault or poaching, or, maybe something worse.[29]

This 'something worse' is, in the course of the story, a sexual assault, as the young Mary Creedon, having left her father at a fair day, is raped and impregnated by a vagrant on her way home. Since the father is unknown, Mary Creedon recasts herself as a new Madonna, with a surprising effect on her society, as 'the boys gave up drinking and fighting and went about quietly and decently, the women kept a watch on their tongues, and the very children playing in the street seemed more gentle and civil'.[30] Christmas Eve is the moment of truth:

28 Signed H. Stuart and C. Salkeld – 'To all artists and writers' *To-morrow* 1:1 (Aug. 1924), 4.
29 Lennox Robinson, 'The Madonna of Slieve Dun', *To-morrow* 1:1 (Aug. 1924), 1.
30 Ibid., 8.

> … towards evening a crowd of people gathered near the cottage, for it was known that the child was coming. The women were in with her, and the men were sitting in the outer room or standing in the doorway, the shepherds from Derry mountain were there, and the sheepdogs lying on the hearth.
>
> It was just a few minutes after midnight when the child was born. They all crowded into the room. They knelt round the bed, the candles shone on Mary and the child.
>
> Mary lay back on the pillow. She opened her eyes, she saw all the people kneeling, she stretched out her arms to them, she laid her hand on the child. –
>
> 'Behold –' she whispered, and she died.
>
> But the child was a girl.[31]

The story ends with a pub scene in Cork, with the 'tramp … telling a story to two young men. "Will you believe me now?" he finished. "Well, you're the boy!" they said, admiringly, and laughed'.[32] Robinson's brutal sense of transgression mutates in Margaret Barrington's short-story 'Colour' to a more sensual consideration of sex and race. Barrington wrote under her maiden name; the wife of Edmund Curtis, a gifted history professor at Trinity College, Dublin, she foresaw the scandal 'Colour' would cause in a university determined during the early years of independence to prove itself an institution in support of the new state. Barrington's marital consideration was short-lived; she ran off with Liam O'Flaherty shortly after. 'Colour' is set in Paris and has four characters: Sally Winter, an American; Alice Greenslade, an Englishwoman; Barrett, a white male; and an unnamed African. The story starts with a discussion of race between the first three, with 'nigger' the frequent descriptive. Barrett is outraged at the acceptability of white women's relationships with black men in France. Sally Winter (whose very name suggests whiteness), loses interest, and finds 'Her mind filled with a vision.'[33]

31 Ibid. 32 Ibid.

33 Margaret Barrington, 'Colour', *To-morrow* 1:1 (Aug. 1924), 8.

> He was walking along the forest path, with the agile grace of a panther. He was naked, save for a coloured loincloth, and her eye dwelt on his barbarous strength, the long limbs, slender and swift in the race, the arms quick to throw the javelin, and the hands, strong, thick, black hands that could throttle and might caress in a clumsy way. She heard his animal-like grunt, smelt the hot smell of his body and suddenly her mind was swept with a curtain of colour, the green of the tropic vegetation, the brown skin, the crimson loincloth, the grey and scarlet parrot and, over all, the blue tropic sky, blue as lapus laluli.[34]

Here is a painter's palette, the explosion of visionary colours bringing to mind an exotic landscape of Gaugin. But the definite colours are binary, the black-and-white of racial separation. Yeats and Robinson had each tried to negotiate difference by force in their sexual strategies, both figuring rape as the mechanism by which woman meets God in art. Barrington's revolution was to develop newly transgressive sex in *To-morrow*, where rape is discarded for erotic physicality. Sally Winter is haunted by a dream:

> that old dream that had haunted her since she could remember. She felt little black hands at her throat, little hands, damp and warm and a little black head nozzling at her breast. A black man-child lay in her arms. She looked round to find someone who would take it, but no one would, and then suddenly she felt that she could not give it up. She held it to her, closer, closer.[35]

Such internal identification draws an immediate external, verbal response; Winter withdraws from reverie to say 'I hate them all, male and female. And now we're on the subject, I want to say that I'm leaving my hotel. One of these niggers, Senegalese Sitris, I suppose they are, arrived yesterday, and I'm not going to stay under the same roof with him any longer than I can help'.[36] She returns, but is now upset; wearing her dark

34 Ibid.
35 Ibid.
36 Ibid.

silk kimono, she is overwhelmed by the image; 'Black against white. The blood rushed to her head'.[37] Deciding to leave Paris,

> She opened her bedroom door. There he stood not two yards away. She noticed his black hands against the white paint of the corridor, and her nostrils dilating caught the heavy negroid smell. Her knees shook as she went back into her room. The heat overwhelmed her, and it was the colourful of Africa.
>
> 'And that fool that said women could not understand these things'.[38]

The story ends, that fool being Barrett, who chided Alice for her refusal to condemn interracial relations. The physicality of the moment is paramount, as hands, noses, knees, conspire surrender to an alterity that compounds everything official Ireland of the 1920s was not – sexual, experimental, hybrid. Racial communion is distant; the African never speaks, and does not exist as an individual, being an impersonal version of the Continent to match the breathing forests of Conrad's *Heart of darkness*. But the deliberate Paris staging of interracial sex in *To-morrow* was the journal's radical departure; the presumed white myths of Christian religion and the classical world, in Robinson and Yeats, are refigured here in a sexual construct that combines the primal, pre-modern (which had already surfaced in the *Klaxon*, whose frontispiece was an African statue in possession of Grace Henry), with the cosmopolitan, that city where strangers meet. Dublin was not yet that cosmopolis. But its fringes practiced experiment in anticipation (and we can see the later genesis of the cabaret in the second and final issue of *To-morrow* when 'The garden', a pseudonymous, and mocking, philosophical dialogue, rebukes a character who could 'see your priests declaiming from the Iliad, or a speech from Macbeth, with evening services from Dr Yeats' poetry'. 'Why not', is the reply, 'Don't you think it would be better than the boredom of fifty Hail Maries?').[39] *To-morrow* could not sustain such internal combustion. The alternate urbanity of sex and the

37 Ibid. 38 Ibid.

39 Sachka, 'The garden', *To-morrow* 2 (Sept. 1924), 2.

modern in Irish literature continued with Samuel Beckett's early contributions to Eugène Jolas' Paris-based quarterly, *transition*. The unnamed central character of Beckett's first published short story, 'Assumption', lives in a constant, isolated, state of anxiety. As prelude, the authorial voice provides a definition of art antithetical to that of Yeats' mode in 'The winding stair'. To Beckett in 'Assumption',

> To avoid the expansion of the commonplace is not enough; the highest art reduces significance in order to obtain that inexplicable bomb-shell perfection. Before no supreme manifestation of Beauty do we proceed comfortably up a staircase of sensation, and sit mildly upon the topmost stair to digest our gratification: such is the pleasure of Prettiness. We are taken up bodily and pitched on the peak of a sheer crag: which is the pain of Beauty.[40]

This perception feeds the rest of the fragment, which narrates, elusively, the entrance of an unnamed woman into the artist's life. She is, at first, an admirer, with 'the usual story, vulgarly told: admiration for his genius, sympathy with his suffering, only a woman could understand'.[41] Her physical description is antithetical to conventional beauty, with eyes 'so deeply set as to be almost cavernous' and cheekbones in 'a misty shadow'.[42] They have sex, and the artist loses, by diminution of his 'essential animality', 'something of the desire to live, something of the unreasonable tenacity with which he shrank from dissolution'.[43]

> Then it happened. While the woman was contemplating the face that she had overlaid with death, she was swept aside by a great storm of sound, shaking the very house with its prolonged, triumphant vehemence, climbing in a dizzy, bubbling scale, until, dispersed, it fused into the breath of the forest and the throbbing cry of the sea.
>
> They found her caressing his wild dead hair.[44]

40 Samuel Beckett, 'Assumption', *transition: an international quarterly for creative experiment* 16–17 (June 1929), 269.

41 Ibid., 270.

42 Ibid.

43 Ibid., 270.

44 Ibid., 271.

This puzzling, obscure imagination of Beckett's may echo a trope that recurred throughout the 1920s in Irish literature: the use of sex – often physically graphic, even violent sex – as a symbol of new social configurations post-independence.

Con Leventhal continued all the while to act as cipher between Europe and Ireland. 'Dada or the Comic Iconoclasts', in the *Dublin Magazine* of October 1935, was a summary of that movement and its influence. Leventhal cast Dada as 'a natural step from the hermetic Mallarmé and the discovery of pure poetry' to a more chaotic sense of being post-world war.[45] He picked up further tremors of Dada in Berlin, in the American *Little Review*, in Gertrude Stein and Wyndham Lewis. Leventhal believed Ireland became involved in the modern movement when

> Through Joyce's *Ulysses* the satiric humour of the age found coherent expression. Paul Morand could only count himself the complete cosmopolitan when he had visited Dublin. In his view buildings are blown up by the natives out of sheer exuberance in order to provide a thrill for the Continental correspondent. Another of Morand's sketches shows an intimate knowledge of the Amazonian nature of the Irish war by making a character confess to having a father who was a general in the Greek army and a mother who was a general in the IRA.[46]

The literal destructiveness of the comic absurd is apparent here, as the Irish revolution translates to a European moment of symbolic deconstruction, from the Atlantic to the Aegean, from Ireland to Greece. The combination of revolutionary political and literary space is most apparent in the next *Dublin Magazine*, in October 1936, in which Leventhal explains his theory of 'Surrealism or literary psycho-therapy'. This issue starts, memorably, with two sets of poems, by Samuel Beckett and Ernie O'Malley. Beckett was by the mid-1930s known, if not liked, for his 1934

45 A.J. Leventhal, 'Dada or the comic iconoclasts', *Dublin Magazine* 10:4 (Oct.–Dec. 1935), 24–5.

46 Ibid., 29.

novella *More pricks than kicks*. Beckett was also a friend of Leventhal and Joyce, and well connected to contemporary vanguard movements in art and literature in Paris. O'Malley was a republican, and author of the memoir *On another man's wound*, which had just been published. To read Beckett and O'Malley together in the *Dublin Magazine* is to sense, for a brief moment, the complex connections between revolution in art and politics as experienced by Irish writers of the immediate post-independence period. First, Beckett's 'Cascando', which starts

is it better abort than be barren

the days after you are gone are so leaden
they will always start dragging too soon
the grapples clawing blindly the bed of want
bringing up the bones the old loves
sockets filled once with eyes like yours
all always is it better too soon than never
the black want splashing their faces
saying again nine days never floated the loved
for nine months
for nine lives[47]

The suggestion of Shakespeare, with the Yorick skull and Ophelian love, afloat nine days, is an echo of the Radicals' Club production of *Macbeth*, with its pillory of Yeatsian grandiosity. Beckett's language here transforms from the relative lyric of the first passage to an allusive scarcity.

terrified again
of not loving
of loving and not you
of being unloved and not by you
of knowing not knowing pretending
pretending

47 Samuel Beckett, 'Cascando', *Dublin Magazine* 11:4 (Oct.–Dec. 1936), 3.

I and all the others that will love you
if they love you[48]

Following this word puzzle, Ernie O'Malley's two poems, 'Grannia to Diarmuid' and 'County Mayo', are conservative in form, though the first of these asks a question unanswerable as Beckett's play.

What did he talk about;
thin spear thrust in the dark
the red glow of a fort
brown stags beneath bare pine
or, a sweet rán well made?

Why should you tell of these
when he could see your eye
glint like a mountain lake,
… when you could hear his voice
change like a gliding wing
that finds its peace in air?[49]

Both poets, following different strategies, consider the effect of sexual impulse on personal psychology, and in so doing continue a dialogue of body and gender in Irish literature that continued from the revolutionary period into post-Independence. It is difficult to reconcile the idea of radical cabaret with a cultural historiography that has defined the immediate post-Civil War period as depressed. But decolonization was a fraught, contested process that continued long after independence and took place in back room clubs and fragmentary journals, as much as in the institutions of Church and State. It is easier to look to Lenten pastorals and Dáil speeches for confirmation of Irelands we already know, of religious and class distinction. In contrast, the ephemeral arts considered here repeatedly failed: failed their audiences, failed their subjects.

48 Ibid., 4.

49 Ernie O'Malley, 'Grannia to Diarmuid', *Dublin Magazine* 11:4 (Oct.–Dec. 1936), 5.

Such failure is necessary. Cabaret is a hybrid form of cultural immediacy; it projects to that space between platform and audience where playgoers become playmakers as they rush the stage by joint agreement. In such fragile consensus exists the possibility of an alternative Ireland in modernity, a first independence of intellect and emotion to express itself in dissent within a new state whose own protocols were still under construction.

'A certain Irish lover of well-bound books'

Sir Edward Sullivan, Bart., 1852–1928

NICOLA GORDON BOWE

IN 2000, IN WHAT WAS DESCRIBED as 'A proposal for the Millennium', the architectural historian and scholar Maurice Craig drew attention to 'probably the most majestic series of bound volumes in the world' that covered the 149 volumes of the seventeenth- and eighteenth-century records of the transactions of each Irish parliamentary session.[2] Miraculously, these large volumes, mostly 'bound in the finest goatskin, generally red in colour ... measuring on average about 21½" x 14½",[3] 'mostly with inlays and tooled with almost incredible lavishness',[4] each different, had survived (bar six) into the twentieth century. That every single one of these 'particularly splendid examples of the bookseller's craft' was 'burned to ashes'[5] when the Public Record Office in

1 Sir Edward Sullivan's description of himself in the paper, 'Decorative bookbinding in Ireland', he delivered in 1911 as 'Brother Bookbinder' (see n. 10).

2 'A proposal for the Millennium – Maurice Craig makes a case for the recreation of some lost Irish treasures', Preface, *Irish Arts Review Yearbook 2000*, 16, pp. 1–2.

3 Joseph McDonnell, *Five hundred years of the art of the book in Ireland – 1500 to the present* (Dublin and London, 1997), p. 16.

4 Maurice Craig, *Irish bookbindings: the Irish Heritage Series 6* (Dublin, 1976), n.p.

5 Inscribed label stuck on to the inside cover of the volume, sold in Sotheby's sale of Sir Edward Sullivan's library, 13 Dec. 1926, records in his hand that: 'The whole of the originals were burned to ashes in the recent Irish Rebellion, the Record Office having been set on fire before the rebels were driven out'.

Dublin was burnt down during the Civil War in 1922, Dr Craig thought was 'pretty generally known',[6] even though 'very few appreciative eyes had ever been cast on them before they perished'.[7] Their destruction, he has written, 'was probably the greatest single artistic loss that Ireland has ever suffered, at least in modern times'.[8] However, much less known was that each of these superlative bindings for the manuscript *Journals* of the Irish House of Lords and House of Commons had been meticulously documented and recorded in photographs and sharply detailed rubbings by Sir Edward Sullivan, Bart. It was, Dr Craig noted, almost as though he 'had some kind of presentiment of their coming destruction. But for him we would have very little idea of what they were like and of what we have lost, for very few Irish bindings approached them in splendour'.[9]

Neither these tragically destroyed masterpieces, nor the many enchanting bindings that Sullivan himself decorated between *c.*1896 and 1926 are much heeded today, except by those with specialist interests in bookbinding or in the history of the respective Georgian and Edwardian periods in which they were made. Similarly, his knowledgeable studies and spirited, still pertinent writings on decorative bookbinding and its design do not receive the attention given to those of his English peers of the same period. The huge folio volume containing his meticulously annotated rubbings and careful yet lively observations of every aspect of the decoration of the bindings he so admired languishes in the National Library of Ireland, along with his detailed notes on the accounts from the Dublin Public Record Office of the binders who made them. He observed in 1911:

> Anyone conversant with the history of art and artistic craftsmanship generally is aware that Ireland in the eighteenth century had attained a notable distinction in connexion with such matters as painting, engraving, miniatures, interior decoration

6 Craig, *Irish bookbinding*, n.p. 7 Ibid.
8 Maurice Craig, 'The Irish parliamentary bindings', *The Book Collector*, II:I (Spring 1953), 24, quoted in McDonnell, *Five hundred years*, p. 16.
9 Ibid. Ironically, the Journals of the Houses of Lords and of Commons 'had been kept in fine condition by their original soft leather outer cases'.

> of houses, wax modelling, and silver chasing; while the Irish medallists, printers ... sculptors and paper-makers had made themselves a name known a long way outside their own shores.[10]

Sullivan reminded English audiences that 'Dublin was then to Ireland what London is to-day to the United Kingdom', and widely renowned for its 'high standard of excellence in many branches of art and craftsmanship'.[11] But, when it came to bookbinding, he regretted that, while 'every nation of Europe' which had excelled in this art had 'been ably exploited',

> You may look through all the volumes that profess to give us the history of Bookbinding in Europe without finding much enlightenment on the subject of Irish work in this branch of artistic expansion.[12]

This, despite the evidence 'that the work which *was* done in this department of the arts is of just as notable a type as anything that was accomplished in the other fields of decorative art'.

Invariably, with only one minor exception, he lamented that

> the best work of such a kind done in Ireland has either been unnoticed by writers on the subject of Bookbinding, or confounded with British work, and for the best of good reasons: the compilers of such records had no knowledge that anything worth

10 Sir Edward Sullivan, Bart., *Decorative bookbinding in Ireland,* no. lxvii, Privately Printed Opuscula issued to Members of the Sette of Odd Volumes, 1911, published privately in a limited edition of 133 copies by the Arden Press, Letchworth, 1914 (see note 19). This important account illustrates some of Sullivan's own bindings along with those by the anonymous masters of the late seventeenth and eighteenth centuries who raised bookbinding 'to something in the nature of a fine art'. When it was reproduced by the Irish Georgian Society (17: 3 & 4) in 1974, it was supplemented by an appendix, which included an 'Alphabetical list of Irish binders' from Charles Ramsden's *Bookbinders of the United Kingdom, 1780–1840* (London, 1954).

11 Sir Edward Sullivan, 'Ornamental bookbinding in Ireland', *The Studio*, 36 (1906), 53.

12 Sullivan, *Decorative bookbinding*, p. 11.

> describing had ever been achieved in that neglected portion of the United Kingdom.[13]

As far as Sullivan was concerned,

> The work that was done in Dublin all through the eighteenth century was in no way inferior, in the matter of artistic feeling, design and general workmanship, to the very best work that was at any time produced in England – and England, as we know, has reason to be proud of its taste and good work in Bookbinding from Tudor days even down the the present.

Furthermore, his observations 'in many of the Irish libraries and in other places throughout the kingdom' had led him to observe that

> There is, besides, in the best Irish binding that distinctive feature, always interesting in studying national tendencies in art, which stamps its leading characteristics as at once distinguishable from similar work done in any other country. In other words, when at its best, it is always original.[14]

Sadly, Sullivan's project in 1904 for the London antiquarian publisher, Bernard Quaritch, to produce a limited edition of 250 copies in a majestic single folio volume of *The Bookbindings/of the/Irish Houses of Parliament/ 1613–1800/ with 50 facsimile reproductions* that he had selected 'for their beauty and originality' did not materialize (Fig. 1). This was to have been 'edited with historical and descriptive particulars by Sir Edward Sullivan Bart.' and 'a summary of the history of bookbinding

13 Ibid. The exception was Mr Weale in his *Catalogue of the bookbindings and rubbings in the National Art Library* (London, 1898) in South Kensington, who referred to 'Some good bindings' being produced 'at Dublin in the eighteenth century, generally in red morocco inlaid with a lozenge-shaped piece of white leather, richly tooled' in one dismissive sentence, and referred erroneously to the high repute of the Dublin stationers' '*vellum bindings*' (Sullivan's outraged italics).

14 Ibid.

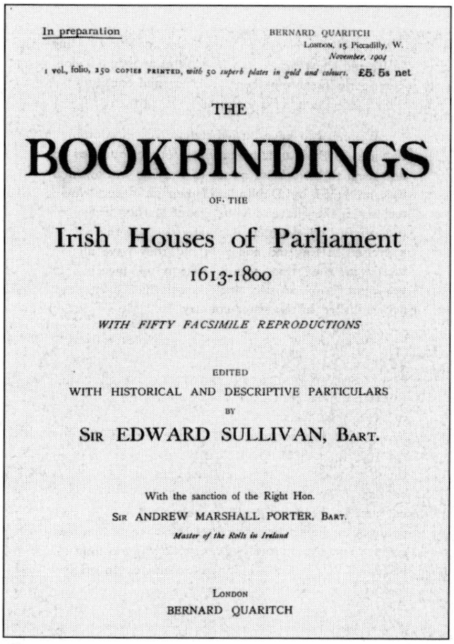

In preparation

BERNARD QUARITCH
LONDON, 15 Piccadilly, W.
November, 1904

1 vol., folio, 250 COPIES PRINTED, *with 50 superb plates in gold and colours.* £5. 5s net

THE

BOOKBINDINGS

OF THE

Irish Houses of Parliament

1613-1800

WITH FIFTY FACSIMILE REPRODUCTIONS

EDITED
WITH HISTORICAL AND DESCRIPTIVE PARTICULARS
BY

SIR EDWARD SULLIVAN, BART.

With the sanction of the Right Hon.
SIR ANDREW MARSHALL PORTER, BART.
Master of the Rolls in Ireland

LONDON
BERNARD QUARITCH

1 Title page of the prospectus for the intended Quaritch publication of *The Bookbindings of the Irish Houses of Parliament, 1613–1800*, edited by Sir Edward Sullivan, Bart. [1904], courtesy of the National Library of Ireland

in Ireland so far as it is known.'[15] The book was advertised with a lavish prospectus. This latter proudly presented as its intended frontispiece a facsimile reproduction of the *Commons Journals* binding for 1747 in sumptuously gold-tooled crimson, black and cream, superbly reproduced by Messrs. W. Griggs & Co. of London; it was accompanied by Sir Edward's four-page introduction (Fig. 2). Apparently, 'support was lacking and the book was never published'.[16] Fifty years later, Maurice Craig paid homage to Sullivan by similarly presenting this by now destroyed masterpiece as the frontispiece for his scholarly publication, *Irish bookbindings, 1600–1800* (London, 1954), while identifying it as the work of

15 The prospectus and catalogue are included in four batches of Sullivan material housed in the National Library of Ireland [hereafter NLI] MS 8389.

16 Fergal McGrath, S.J., *Father John Sullivan, S.J.* (Dublin, 1941), p. 18.

2 *The Commons Journals, 1747*, featured in the prospectus for Sullivan's proposed Quaritch book, courtesy of the National Library of Ireland

Parliamentary Binder A. His ground-breaking study was partly dedicated to Sullivan, whose manuscript notes, rubbings and publications he affirmed as essential for any study of the subject.

While Sullivan's seminal documentation was similarly acknowledged by Joseph McDonnell in the publication accompanying his exhibition of 'Five hundred years of the art of the book' at the National Gallery of Ireland in 1997, Sullivan was actually represented in the exhibition by a blue goatskin binding he himself had 'finished' on a small sixteenth-century Venetian volume (Fig. 3).[17] Tooled in gold, with red and citron onlays, its floral and foliate decorations and central oval lozenge on the cover are simplified contemporary homages to the anonymous eighteenth century binders he so admired, yet his arrangement of small pink onlaid hearts, tudor roses, profuse dots, vertical lily stems, and his enigmatic signature 'E.S. AURIFEX' place his design firmly within an Arts and Crafts idiom. Dr McDonnell sited Sullivan's own work in the context of the widely influential late nineteenth century revival of bookbinding in England led by T.J. Cobden-Sanderson, a founding member of the English Arts and Craft Exhibition Society, whose bindings he greatly admired and who, like Sullivan, was trained as a barrister.[18] Sullivan's importance in the Irish Arts and Crafts movement had already

17 The book was Alberto da Castello's *Rosario Venet* (Venice 1579), presented by Sullivan to the O'Brien Library at the Jesuit Milltown Park in Sept. 1907. I am grateful to the late Father Fleury, S.J. and the current librarian, Trish Quigley, for showing me the Sullivan material in Milltown library. Although Sullivan described himself as a bookbinder, properly speaking he was a 'finisher' rather than a 'forwarder' of bindings, which means that he tooled and gilded the leather, rather than putting the leather on the boards. McDonnell, *Five hundred years*, p. 164 writes that 'finishing' is 'the art of decorating the covers with gold-tooling and inlaying or (more likely) onlaying different coloured leathers (or papers).'

18 Cobden-Sanderson practised as a barrister in London from 1871 to 1883, before taking up bookbinding at the suggestion of William Morris's wife, Janey; Sullivan was called to both the English and Irish Bars, but never practised. For Cobden-Sanderson's role in the formation of the English Arts and Crafts Movement, see Gillian Naylor, *The Arts and Crafts Movement: a study of its sources, ideals and influence on design theory* (London, 1980), p. 123; and for his work, M. Tidcombe, *The bookbindings of T.J. Cobden-Sanderson* (London, 1984). For the influence of Cobden-Sanderson on the Arts and Crafts Movement in Ireland, see Nicola Gordon Bowe, 'The book in the Irish Arts and Crafts Movement' in Clare Hutton (ed.), *The Irish book in the twentieth century* (Dublin, 2004), pp. 16–35.

3 Lower cover of binding finished by Sir Edward Sullivan, blue goatskin, gold tooled, with red and citron onlays, *c*.1905, signed 'E.S. Aurifex', on Alberto da Castello, OP, *Rosario Venet* (Venice, 1579), courtesy of the Jesuit Library, Milltown Park, Dublin

been signalled in an exhibition of 'The Dublin Arts and Crafts Movement, 1885–1930' at the 1985 Edinburgh Festival, and pursued in Paul Larmour's book, *The Arts and Crafts Movement in Ireland* (Belfast, 1992). It was further examined in the book, *The Arts and Crafts Movements in Dublin and Edinburgh, 1886–1930*, which chronicled both his important role in the formative years of the Arts and Crafts Society of Ireland in 1894, and his key published writings on design and described four bindings he had decorated between *c.*1901 and 1924.[19] It is this aspect of his contribution to the history of Irish bookbinding that will mainly be considered here.

Sullivan's own bindings are not generally appreciated by those who are versed in the stunning complexities of what he described as 'the amazing versatility and manifest individuality'[20] of the eighteenth-century Irish masters. Until recently, his work has tended to be dismissed as derivative even if the execution of their tooling has been acknowledged as 'admirable', along with his choice of materials.[21] A critic who acknowl-

19 Nicola Gordon Bowe and Elizabeth S. Cumming, *The Arts and Crafts Movements in Dublin and Edinburgh, 1885–1925* (Dublin, 1998), the publication in book form comprises the twin catalogues which accompanied the exhibition (Edinburgh, 1985). Sullivan's bindings featured in the *Dublin Arts and Crafts Movement, 1885–1930* exhibition catalogue, nos. 143–6, were: a brown and green morocco Writing Case of 1909, onlaid and tooled, signed with his 'ES' monogram; an onlaid and tooled green morocco binding on Robert Anning Bell's illustrated Shakespeare's *The Tempest* (1901); an onlaid and gold tooled pink calfskin album for R.R. Holmes' *The life of Queen Victoria* boldly stamped '19 E.S. 00'; a miniature inlaid, tooled binding for Kate Greenaway's 1893 *Almanac*, printed by Edmund Evans; Sir Edward Fitzgerald's *Rubaiyat of Omar Khayyam* (1924), with an elaborately tooled, onlaid and painted red morocco front and back covers and spine, signed with a distinctive small 'S' monogram and 'E.S AURIFEX'. The consequent book also featured: 'Dublin verses by members of Trinity College', published in 1895 by Elkin Matthews and Hodges, Figgis & Co., whose green morocco covers display geometrical enclosures of onlaid, painted and tooled stamps, including shamrocks and Celtic interlace; and Sullivan's own account, *Decorative bookbinding in Ireland*. This was read as a paper before Ye Sette of Odd Volumes on 28 February 1911 at the Hotel Capitol (Oddenino's) by 'Sir Edward Sullivan, Bart., bookbinder to Ye Sette' and subsequently issued as no. lxvii of privately printed opuscula to members of The Sette of Odd Volumes as Opusculum Irvii, 1914, in a limited edition of 133 copies, 'imprynted for private circulation only' in small octavo by the Arden Press at Letchworth (1914).

20 Sullivan, *Decorative bookbinding*, p. 16

21 Dr Maurice Craig, letter, 30 Apr. 1985.

edges that, 'At its best, Sullivan's work rivals that of far more famous craftsmen of his day, both for design and quality of execution', also writes that 'At its worst, Sullivan's workmanship can be shown to be indifferent, imprecise and crooked. His lettering is so carelessly executed as to make embarrassingly clear the amateur nature of his work'. The same critic considers that his lettering, in particular, 'suffers from poor arrangement, spacing, and the incongruous selection of inappropriate sans-serif type styles. When examined closely Sullivan's decorative tooling sometimes appears busy and overly fussy'.[22] Such judgments may be the result of inadequately heeding his unique synthesis of current Arts and Crafts practice, epitomized in the free-form bindings of Cobden-Sanderson and his followers, with the clear influence of the brilliantly idiosyncratic decorative flourishes of the eighteenth century Irish binders 'where repeated use of light-coloured leather onlays and a profusion of small tools are geometrically arranged'.[23] They may also stem from seeing poor black-and-white reproductions of his work illustrated in articles he wrote, rather than a representative range of his most successful richly coloured, distinctively decorated leather bindings. The difficulty is that they relatively rarely appear on the market, despite the fact that it has been estimated that 'over the span of little more than a quarter-century Sullivan decorated more than 200 volumes', perhaps as many as double this number.[24] When twenty-three diverse bindings, 'perhaps the largest [collection] in private hands', turned up at auction in 1988, Sotheby's celebrated this unique event by dedicating their catalogue cover to a delightfully romantic close-up detail of a Norse ship set in Celtic whorls and heart-strewn lilies decorating his binding (1920) on Hilaire Belloc's translation of J. Bédier's *The romance of Tristam and Iseult*.[25] In 1941, his brother's

22 Thomas V. Lange, 'The decorated bookbindings of Sir Edward Sullivan, Bart.', *Long Room* 40 (1995), 47. 23 Ibid.

24 Ibid. The antiquarian booksellers, Maggs Bros. Ltd in London and De Búrca Rare Books in Dublin have both held important Sullivan bindings, and I am grateful to each for information. Lange records that, in the Huntington albums, 'only two prices appear for books Sullivan sold: one to Maggs for £7 in 1901, the other to Bain for £6 in 1900' (p. 46).

25 *Maps, antiquarian and modern books and original illustrations*, sold by Sotheby's at Billingshurst, West Sussex, 27 Apr. 1988, lot 2283, p. 113. The collection was given over a period of fifteen years to one of his closest friends, Nina Barker, to whom a

biographer, Fergal McGrath, SJ, having consulted the Director of Sotheby's and Colm O'Lochlainn, eminent binder and printer of the Three Candles Press, wrote that 'Specimens of Sir Edward Sullivan's own work as a bookbinder are treasured by connoisseurs'.[26] The fact that Sullivan's bindings are all 'finished', which means that 'he applied the tooled decoration to books already bound by others',[27] and never seems to have 'forwarded' the entire binding from its beginning, has been held against him, even though his work is to be found in distinguished public and private libraries and collections on both sides of the Atlantic.[28]

The first full-length study of 'The decorated bookbindings of Sir Edward Sullivan, Bart.' appeared in 1995. Written by Thomas V. Lange, Curator of Early Printed Books and Bindings at the Huntington Library in California, it focused on three 're-used Victorian folio volumes' acquired by the Huntington in 1936 which contain 'pencil rubbings of 208 works finished by Sullivan', made between April 1896 and June 1926.[29] These albums remain 'unpublished in the literature of bookbinding'.[30] Before then, the only references to Sullivan's work were to be found in his own writings in the *Journal and Proceeding of the Arts and Crafts Society of Ireland*, in *The Studio* magazine, in his account of

number of his bindings were dedicated; some are inscribed with her gold-tooled initials, or with an onlaid, gold-tooled bee. I am grateful to Robert Ansell of Sotheby's Book Department for advice in this respect.

26 McGrath, *Father John Sullivan, S.J.*, p. 18.

27 Lange, 'The decorated bookbindings of Sir Edward Sullivan', 41, who describes him as 'a staunch advocate of the history of Irish binding and an enthusiastic bookman'.

28 Notably, the British Library, Trinity College, Dublin Library, the Huntington Library, the late Sir Paul Getty's library at Wormsley, the late Earl of Iveagh's library at Farmleigh, the Victoria & Albert Museum library, the Aidan Heavey collection at Athlone Public Library. Lange also mentions the Folger Shakespeare Library, the Library of the French Senate and the Chapter Library of Westminster Abbey.

29 Lange, 'The decorated bookbindings of Sir Edward Sullivan', 41–7. Lange presumed the rubbings, which appeared 'to have been assembled at different times', had been made by Sullivan himself. Like his studies of seventeenth-and eighteenth-century bindings, they are 'all executed on the same gossamer tissue paper pasted to album leaves', and are punctiliously annotated, described and documented. I am grateful to the late Vincent Kinane for discussions in this respect.

30 Lange, 'The decorated bookbindings of Sir Edward Sullivan', 46.

'Decorative bookbinding in Ireland',[31] in the favourable critical appraisals of his work by his peers in the early years of the Arts and Crafts Society of Ireland, in a biography of his brother (who had become a Jesuit in 1900),[32] and in a specially printed posthumous tribute to him. Entitled *Brother Bookbinder (Sir Edward Sullivan, Bart.)*, this was compiled by Ivor Stewart-Liberty of the Sette of Odd Volumes in London, with which Sullivan had become deeply involved since joining it in 1903, and to whose members the small ottavo volume was privately circulated as Opusculum lxxxvii in February 1929.[33]

Lange's ground-breaking article is crucial for a number of reasons. Most of all, it reveals the importance of the Huntington volumes in dating exactly when Sullivan executed a great many, although not all, of his bindings. In the albums he noted 'detailed pencil descriptions of colours, materials, inlays (actually onlays), designs, the forwarders (when commissioned), the date completed, and often an identification of for whom the work was done or to whom it was given'. Without this reference, except in the few cases where the dates are gold-tooled on as part of the binding,[34] dating Sullivan's bindings can only be guesswork related to stylistic analysis, tools and inscribed dedications. Now Lange could safe-

31 Sir Edward Sullivan, Bart., 'Early Irish printers and their work', *Journal and Proceedings of the Arts and Crafts Society of Ireland*, 1:1 (1896), 43–6; Sullivan, 'Irish bookbinding, and how it may be improved', *Journal and Proceedings of the Arts and Crafts Society of Ireland* [hereafter *Journal and Proceedings*], 1:1 (1896), Appendix, pp. 57–9; Sullivan, 'The Queen's album', *Journal and Proceedings*, 1:2 (1898), 167–9; Sullivan, 'Design in gold–tooled bookbinding', *The Studio*, 33 (1905), 34–9; Sullivan, 'Ornamental bookbinding in Ireland in the eighteenth century', *The Studio*, 36 (1906), 52–9; Sullivan, *Decorative bookbinding*.

32 McGrath, *Father John Sullivan S.J.*

33 This was printed in small ottavo by Alabaster, Passmore and Sons Ltd in Cannon Street, London in a limited edition of 101 copies. The author was a nephew of the late Sir Arthur Liberty, D.L., J.P., who was known as 'Brother Craftsman' in the Sette and was an old friend of Sullivan's.

34 For example, his signature, emulating that of Cobden-Sanderson, '19 E.S 00', gold-stamped on the doublure of his unaccustomedly restrained decoration on the Royal Librarian Richard P. Holmes' 1897 tome, *The life of Queen Victoria*; and 'Christmas 1909' along with 'Write till your Ink be Dry' Shaks' and his monogram 'ES' on a richly tooled writing case (both Heavey collection, Athlone). My grateful appreciation to Aidan Heavey for making his fine collection so accessible to me.

4 Sir Edward Sullivan's colophon as Brother Bookbinder for Ye Sette of Odd Volumes

ly state that 'The bindings decorated by Sullivan fall into three distinct categories: new bindings he commissioned and decorated, publishers' bindings he decorated, and old bindings he acquired and decorated'. Not only this, these albums recorded the identities of the binders Sullivan had commissioned to cover sixty-three out of the 196 completed books and albums whose rubbings he had made, some in Dublin, others in England, Leipzig and Florence. Another sixty, Lange reckoned, could 'reasonably be expected to have come to Sullivan in their publisher's bindings', notably 'dated annual diaries, Bibles, prayer books, dictionaries' and sixty-one of the 'notoriously difficult to tool' miniature volumes (under 3 inches high) he seems to have greatly enjoyed decorating.[35] Lange also refers to Sullivan's willingness, unlike Cobden-Sanderson, to decorate 'non-book items', such as the blotters, jewel box, manicure case and mirror case recorded in the Huntington albums, and the slip-on book cover, stamp book holder, wallet, and inkpot and quill holder that turned up in the Barker collection sold by Sotheby's in 1988. Sometimes these employ the same tools as larger surfaces, but often they entailed his having to make tools (or *petits fers*, as he referred to them)[36] bearing smaller versions of his accustomed vocabulary of designs.

35 I am grateful to Bryan D. Maggs (letter 7 Mar. 2005) for drawing my attention to a miniature library in the late Sir Paul Getty's library at Wormsley 'in a special bookcase made by Sullivan and all 33 tiny volumes decorated by him'.

36 Sir Edward Sullivan, 'Design in gold-tooled bookbinding', p. 38.

5 Upper cover of binding finished by Sir Edward Sullivan, blue goatskin, gold tooled, with maroon and orange onlays, 1914, signed 'E.S. Aurifex', also 'S' monogram on the inside board, on *Huon of Bordeaux: done into English by Sir John Bourchier, Lord Berners: and now retold by Robert Steele* (London, 1895), courtesy of the Benjamin Iveagh Library under the Guinness family

5a Lower cover of previous binding

Lange considers that Sullivan was principally concerned to record his designs, taking 'particular pains to document the areas he had decorated: the front cover, spine, turn-in and/or doublure, the location of his monogram and other details such as inscriptions'.[37] That he almost invariably signs his work with his monogram 'ES', his entwined initials, an angular 'S' monogram, or his assumed signature of 'E.S. Aurifex' (that is, worker in gold), unlike the eighteenth-century master-craftsmen he so admired, was partly for identification purposes but also because of the Arts and Crafts ideological insistence that the maker's name be clearly inscribed on his or her artefact. He obviously enjoyed dedicating his work to people he liked or admired, rarely selling but usually giving away or officially presenting his artwork. A typical example in the Iveagh collection at Farmleigh, Dublin, floats a flowery, leafy and dotted scalloped lozenge bordering a central vesica beside two long-stemmed lilies, the entwined initials 'DH', and a quotation from Shakespeare's *As you like it*: 'GOOD-DAY, AND HAPPINESS, DEAR ROSALIND'. JUNE.1914' on a blue morocco ground. Around it is a symmetrical band of Sullivan's signature pimpernels, hearts and dots demarcated by gouges, small rings and a roll of single dots and his distinctive tiny stamp, 'E.S. AURIFEX' (Figs 5 & 5a).[38]

He also took pride in decorating bindings on his own writings and his translations of others' writings. These include six different copies of [his edited] Guazzo's *Civile conversation* in the Huntington rubbings and, elsewhere, at least two versions of his translation into English of 'Hell' from *The comedy of Dante Alighieri* published by Elliot Stock in 1893 (Figs 6 & 6a). One, inscribed to his sister Annie, is richly tooled, dramatically gouged, onlaid and painted on contemporary brown morocco within framing rolls on its upper cover, with a more restrained lower

37 Lange, 'The decorated bookbindings of Sir Edward Sullivan', 42. It should be noted that Sullivan often seems to have treated the decoration of the back, or lower, cover quite differently from the front, but no less distinctly.

38 The formalized lily is repeated in twenty vesica enclosures and single dot runs on the lower cover of *Huon of Bordeaux done into English by Sir John Bourchier, Lord Berners; and now retold by Robert Steele* (London, 1895). Sullivan's small 'S' with bubbles is stamped on the book's upper and lower doublures. I am grateful to Julia Cummins, Librarian at Farmleigh, for showing me this and other Sullivan bindings collected by the late Earl of Iveagh.

cover, whose centrally decorated vesica bears his entwined initials. The other, seemingly much later version, bears a similar entwined monogram on its upper cover, within a larger, looser, scalloped centrepiece, and more compartmentalized upper cover; the lower cover declares an upper-case, gold-tooled inscription: 'TO WISH YOU HAPPY DAYS' above the large, onlaid leather initials of Nina Barker floating in a scalloped floral and foliate centrepiece and, below, the date '17.3.1922'; along the base linear roll in the same clear upper case, hovers a single initial 'S' pierced by what look like two tiny bubbles forming the letter 'e'. Sullivan's design recalls one of the architectural features McDonnell refers to as being 'of great consequence for the Irish Parliamentary binders of the early eighteenth century', whose effect 'owes a good deal' to the designs of Persian carpets: 'an inner rectangle with onlaid centre and corner pieces of contrasting colours'.[39] The binding Sullivan decorated for his own account of *Decorative bookbinding in Ireland* (1914) is similarly described: 'Contemporary brown morocco, gilt tooled foliate border, gilt tooled floral corner ornaments surrounding a central onlaid lozenge of green morocco tooled and onlaid'.[40] His profound knowledge of the history of bookbinding is witnessed not only by his writing and studies but also by his readiness to decorate or re-decorate old bindings, seventeen of which Lange has counted in the Huntington albums, some by unknown binders, others named. Lange comments: 'In binding circles it is unusual to dare to seek earlier bindings to decorate', yet 'Sullivan's decorating of old and plain bindings shows an innovative flair and appreciation of the original style of binding, without the slightest intent to deceive'.[41]

Sullivan was fortunate in that he grew up in Dublin amidst 'one of

39 McDonnell, *Five hundred years*, p. 13. Close analysis of Sullivan's designs and his tools shows his knowledge of and indebtedness to the Irish binders.

40 Sotheby's, *Maps, antiquarian and modern books* (1988) lot 2303, p. 116.

41 Lange, 'The decorated bookbindings of Sir Edward Sullivan', 47. This is exemplified by one of his earliest bindings, which he presented to the Library of his *alma mater*, Trinity College, Dublin in 1892: *The 'Breeches' Bible* (London, 1599), whose faded green morocco upper cover is onlaid with tan and red, tooled respectively with birds and shields. Its inscription is signed 'Eduardus Sullivan, Aurifex et Inventor'. I am indebted to Charles Benson, Keeper of Early Printed Books at Trinity College Dublin Library.

the most valuable private libraries in the Kingdom'[42] collected by his father, an 'ardent and lifelong bibliophile',[43] classicist, linguist, distinguished statesman, brilliant barrister, Liberal MP and immensely powerful lord chancellor of Ireland. When young Edward was eleven, he moved with his parents, older sister and three younger brothers from an eighteenth-century house in Eccles Street to Fitzwilliam Place, no. 30 and then no.32; there he lived until 1910, first with his mother and sister after his father's death in 1885, then with his sister after their mother's death in 1898. He attended Portora Royal School, Enniskillen with his similarly scholarly Dublin neighbour, Oscar Wilde, before proceeding to win classical honours at Trinity College, Dublin and being called to the bar at King's Inns, Dublin (1879) and the Middle Temple (1888). After his father's death, he stood unsuccessfully for parliament, in 1886 and 1892, before deciding to devote his life to literature and the study and practice of bookbinding. Key factors in this decision seem to have been the 'Exhibition of bookbinding given by the Burlington Fine Arts Club in the year 1891' in London, and the example of 'the practical pioneer in bookbinding ... Mr Cobden-Sanderson, and his ornamental bindings in a style altogether his own ... not yet ... surpassed in any country'.[44] He wanted his own bindings to manifest Cobden-Sanderson's belief 'that artistic results in the way of design are more readily to be obtained by the use of a few tools than by using a large number', thus maintaining 'the direct simplicity of the main design', and avoiding what 'Ruskin would call "a glittering vacillation of undisciplined enchantment"' in his *Seven lamps of architecture*.[45] In 1893, Cobden-Sanderson,

42 Part of this, when sold at auction in 1890, reached £11,000: entry on Sir Edward Sullivan, 1st Bt. (1822–85) in Sir L. Stephens and Sir S. Lee (eds), *Dictionary of national biography*, xix (Oxford, 1922).

43 C.L. Falkiner, 'Sullivan, Sir Edward, first baronet (1822–1885)', *Oxford dictionary of national biography* (Oxford, 2004).

44 Sullivan, 'Design in gold-tooled bookbinding', 37. Cobden-Sanderson, twelve years Sullivan's senior, had trained in a traditional London bookbinding workshop before setting up on his own in 1884. By 1893 he was able to set up his Doves Bindery near his friend William Morris in Hammersmith. Although he only worked as a binder for about ten years, his influence was immense, through his work, his teaching and his writing.

45 Sullivan, 'Design in gold-tooled bookbinding', 38.

in true Morrisian mode, would lament the current 'exaggeration of technical skill' and mechanical production; he avowed that 'the life of bookbinding is in the dainty mutation of its mutable elements':

> It is individual; it is instinct with the hand of him who made it; it is pleasant to feel, to handle, and to see; it is the original work of an original mind working in freedom simultaneously with hand and heart and brain to produce a thing of use, which all time shall agree ever more and more also to call 'a thing of beauty'.[46]

One of Sullivan's earliest bindings, which he presented to the eminent Trinity classicist and provost, Dr J.P. Mahaffy in 1891, shows his tentative adaptation of the specifically Irish eighteenth-century idiom of central and corner fields of decoration, scalloped with gouges and filled with a plethora of tiny dots, arabesques of floral and foliate stamps, acorns and stars.[47] There can be no doubt that his detailed studies of the varied designs used in the Irish Parliamentary bindings informed his own choice, design and use of tools. His observations include references to: fans, gold, border and double peacock rolls, painting, dotting, corner arrangements, central lozenges, the colours of the morocco leather used, the use of vellum, lettering, fillets, scrolls, strapwork, radiating foliated curves, the division and inlaying of centrepieces, compartments, imitation *pointillé*, inlaid flowers, stars and birds.[48] By 1893, he had made enough 'fine leather bindings' to exhibit at the English Arts and Crafts Exhibition Society's fourth exhibition in London.

The following year saw the foundation of the Arts and Crafts Society of Ireland by Dermot Bourke, seventh earl of Mayo, a disciple of William Morris, antiquarian, collector and champion of Irish cabinetmaking.

46 T.J. Cobden-Sanderson, 'Bookbinding' in *Arts and Crafts Essays by Members of the Arts and Crafts Exhibition Society* (London, 1893), pp. 147–8; he also emphasizes that 'there should be no distinction between "finisher" and "forwarder"' (p. 144).

47 Bound in dark blue calf on 'La Liturgie c'est a dire le formulaire des priéres publiques de l'Eglise d'Irlande' (Dublin, 1666), on loan to Trinity College Dublin Library.

48 NLI MSS 8389(2) and 8389(3), where he notes in amazement: 'Flowers picked out with *black paint*'.

6 Upper cover of binding finished by Sir Edward Sullivan, tan morocco, gold tooled, with green, pink and black onlays, *c*.1894, signed by Sullivan on the inside board, on 'Hell', first volume of *The comedy of Dante Alighieri* rendered into English by Sir Edward Sullivan, Bart. (London, 1893), courtesy of Eamonn de Búrca, Antiquarian Bookseller, Dublin

6a Lower cover of previous binding

Lord Mayo's express aim was to encourage 'Contemporary original work in decorative design and handicraft' and to try and improve artistic levels of craftsmanship which had been deteriorating, even disappearing, in Ireland since the eighteenth century. For the Society's first exhibition in 1895, in which the names of all artists, designers, workmen and firms exhibiting were to be clearly identified on their work so as to 'make the workman less of a machine producing many objects from one pattern',[49] Mayo had enlisted William Morris' support and a loan exhibition from their English namesake Society. Before the exhibition opened, Sullivan was part of a distinguished deputation from Dublin which accompanied Lord Mayo to Belfast to drum up support for the Society. He was a founding member and guarantor of the Society and served on its Executive Committee as well as its Amusements, General Management and Retrospective Irish Arts and Crafts Section Committees. He could not possibly have been more involved. He also exhibited a case of twenty-two bookbindings and a letter-paper case that he had 'designed, inlaid and tooled', clearly giving the names of the binders who had forwarded each volume (mostly Birdsall of Northampton, although three were from Roger de Coverly with whom Cobden-Sanderson had trained). His work for the exhibition's substantial Retrospective Section, selected 'to give a general aperçu of Irish Art-craftsmanship at a period when design was more or less homogeneous', led to a splendid display of over a hundred of the finest seventeenth-, eighteenth- and nineteenth-century 'Irish bookbindings and printed books', a number of which he lent from his own collection. This magnificent display concluded with an English loan collection from the Arts and Crafts Exhibition Society, arranged by Sydney Cockerell (Morris' librarian and secretary), Halsey Ricardo, architect, and Cobden-Sanderson, who was represented by seven books he had designed, bound, and tooled and four he had designed and his Doves Bindery had bound and tooled. Sullivan's bindings were praised in William Hunt's published Report on the Irish exhibition: 'Here we have in many instances excellence of design coupled with sound workman-

49 *Journal and Proceedings*, 1:1 (1896), 4.

7 Upper cover of binding finished by Sir Edward Sullivan, red crushed morocco, gold tooled and onlaid, signed 'E.S. Aurifex', on Pompilius, *Carmen Pompilii de Triumpho Granatensi* (Rome, 1490). Inscribed: 'Presented to the Library, Trinity College, Dublin by Sir Edward Sullivan, Bart., April 1907', lower cover tooled with lozenge inscribed: 'Almae Matri T.C.D./ E.S. Auri/fex'. Illustrated in Sullivan's 1905 *Studio* article on 'Design in gold-tooled bookbinding', courtesy of the Board of Trinity College, Dublin

ship. Time may mellow these volumes, but the design and workmanship will remain'.[50]

The following year, the Society's first *Journal* published two documents in connection with their first exhibition: Lord Mayo's on 'Cabinet-maker's work' and Sullivan's on 'Irish bookbinding, and how it may be improved'. This was an exhortation to modern craftsmen to emulate the originality, art and craftsmanship of the Irish binders of 150 years ago, to shake free of 'worn-out conceptions', to experiment with the endless 'combinations of ornament ... which may be effected with, say a flower, a bud, a leaf, a dot and a couple of gouges',[51] or the 'many exquisite forms of Celtic design supplied by our national manuscripts'. What was needed was the artistic spirit and originality, he considered, was only to be found in the 'modern books ... of Mr. Cobden-Sanderson of London' with their 'unpretending work, plain and perfect in its simplicity', true to the nature of their materials.[52] In 1897, Sullivan gave one of the Society's lectures in the Royal University, on 'Book-binding', 'intended for practical workers and employers' and judged the artistic leatherwork in the Royal Dublin Society's annual exhibition of craftsmanship.

There is, however, no mention of his playing any role whatsoever in the ACSI's second exhibition held in Dublin in 1899, only a reference at the end of the catalogue to 'The Queen's Album', 'produced under the supervision of the Arts and Crafts Society of Ireland', forwarded in 'pale green crushed morocco' by Marcus Ward & Co. of Belfast, and decorated by Sullivan. Inside were one hundred platinotype views of Irish scenery, 'tastefully mounted' by the Irish Tourist Association for presentation to the Queen via HRH, the Duke of York on a royal visit.[53]

50 *Journal and Proceedings*, 23. Sullivan also contributed a learned article on 'Early Irish printers and their work' to the same issue of the *Journal and Proceedings*, 43–6).

51 Mirjam Foot in vol. 2 of her *Catalogue of North-European bindings* (London, 1983) has traced the whereabouts of other volumes displaying the same tools as those used by Sullivan on the British Library's Henry Davis Gift collection (nos. 267–270). I am grateful to Philippa Marks, Keeper of Bookbindings, British Library, for drawing my attention to this and for informing me about the Library's Sullivan bindings.

52 Appendix to *Journal and Proceedings* (1896), 57–9.

53 'The Queen's album' by Sir Edward Sullivan, Bart.,' *Journal and Proceedings,* 1:2 (1898), 167–9.

It had been exhibited by the ACSI at the Inaugural Art Exhibition of the People's Palace in Glasgow in January 1898, after which Sullivan wrote a short illustrated account of it for the second issue of the ACSI *Journal* later that year. Following English Arts and Crafts principles that his finishing should harmonize with the contents of the book, he had used 'purely Celtic' ornamention on the exterior, 'the first attempt to work out a Celtic pattern of any complexity in gold tooling on the cover of a book', as far as he was aware. His source for what looks like a unique treatment, using small leather inlays in crimson and shades of green, red and orange, was 'the round window of the ancient church at Rahan', and some of his tools had been cut from designs by Richard Caulfeild Orpen, architect. The dark red morocco, liberally inlaid doublures were entirely Sullivan's own design, revealing a border framework and centrepiece of white and red roses symbolizing the royal houses of York and Lancaster, with his signature corner and border groups of alternating long-stemmed orange and white lilies.

Sullivan was back again on the ACSI's Executive Committee for their third exhibition in 1904, although not represented, probably because he was concentrating on his ill-fated Quaritch volume, preparing a study on the authorship of Shakespeare's plays, *Verulamania: some observations on the making of a modern mystery* (1904), editing the eighteenth-century *Memoirs of Buck Whaley* (1906) and spending more time in London, based at the Reform Club. By 1906, he was listed as living in lodgings in London's west end, as well as in Fitzwilliam Place. The years 1905 and 1906 saw the publication of his two important articles on bookbinding in the progressive applied arts magazine, *The Studio*. The first, 'Design in Gold-tooled bookbinding', is exclusively illustrated by five of his own bindings, each diverse yet immediately recognizable through his tools and scalloped centrepieces; one, on 'The tale of the Emperor Coustans and of Over Sea' (1894), Icelandic sagas translated by William Morris, had been singled out in the Report of the ACSI's first exhibition.[54] His

54 This volume, printed by Morris's Kelmscott Press and now in the Huntington Library's collection, was selected to represent Sullivan at the Los Angeles County Museum of Art's exhibition, *The Arts and Crafts Movement in Europe and America, 1880–1920: design for the modern world* (2004).

aim in this article is to define what makes what he calls a 'bibliopegic jewel': 'choice craftsmanship'; nothing but the finest materials used with skill, revealing the very 'soul of the artificer, outlined in graceful curve or clustered loveliness'; and the choice of 'some simpler form' preferably selected for its 'design and artistic repose', in keeping with the spirit of the time. Citing Cobden-Sanderson as unique in his original, contemporary expression and historical understanding of bookbinding, Sullivan emphasizes the importance of sticking to the use of a few tools in order to achieve artistic design results, which do not necessarily have to reflect 'the period at which the book was composed or printed'. He proceeds to refer to his own 'mosaic work', whose decorative effect he considered 'largely due to a harmonious blending of inlaid, or superimposed, pieces of coloured leather set in contrast with the differently coloured ground in which they are ... imbedded'.[55] He concludes with three further observations on design: that the doublures (or inner linings of the covers) are the only areas where a 'modified extravagance' is acceptable to contemporary taste; the importance of maintaining a design relationship between the upper and lower covers, even if (as they often were in his case) different from each other; and the danger of the exterior decoration overdoing the connection with its contents as, he considered, the 'modern French' binders were inclined to do.

His recommendation that 'some obvious and easily intelligible connection can be established, by form of tool or general scheme of decoration, between the outside and the contents of the volume' can be seen in his charming upper cover on G. Baruffaldi's *La Tabaccheide* (1714), where four eighteenth-century smoking pipes are symmetrically gold-tooled on the tobacco-coloured morocco binding in each corner between the decorative centrepiece and borders (Fig. 8).[56] His aspiration was that his own work might be seen to display that 'power to impress upon his creations an individuality of treatment as well of design'. Nowhere was this so evident, he thought, as in the 'remarkable originality and beauty' of Ireland's eighteenth-century bookbindings, the subject of his 1906

55 Sullivan, 'Design in gold-tooled bookbinding', 38.

56 I am indebted to Eamonn de Búrca for showing me this and other Sullivan bindings.

Studio article; not that anyone could match 'the magnificence and bewildering variety', 'the intricate and luxurious mosaic designs', the exuberant colours, inventive inlays and tooling, and 'fine effects of harmony' achieved by the unnamed Parliamentary Binders in 'the very acme of the bookbinder's art'.[57] Sullivan's own bindings continued to refer to and pay homage to them over the next twenty or so years, only occasionally deviating from their characteristic use of centrepieces to produce an all-over design of great charm. Notable is his bee-strewn floral decoration gold-tooled, onlaid and painted on *The poems of Edmund Spenser* (1886) in red morocco, surely inspired by the 'simple, yet pretty pattern' of the small Prayer Book binding of 1747 he illustrates in his 1906 article.[58] When The Sette of Odd Volumes published his study of *Decorative bookbinding in Ireland* in 1914, three of his own bindings were reproduced along with those for two Commons' *Journal*s and one Lords' *Journal* in the ultimate homage. Modestly, he concluded with what may serve as his own epitaph:

> Towards the end of the nineteenth century a certain Irish lover of well-bound books took it into his head to study in a practical way the interesting processes by which decorative bookbindings are produced. He has worked at it intermittently ever since in his own amateurish way, endeavouring all the time to keep before his eyes that artistic standard without which no good work can ever be accomplished.

Despite his lifelong love of bookbinding, Sir Edward Sullivan is now best remembered for his pioneering and scholarly study of *The Book of Kells*, published in a folio volume by the Studio Ltd in London in 1914. His book was the first to reproduce in colour of any quality 'one of the most interesting and beautiful manuscripts which has yet come from the hands of man', using the latest technology to render it as close to the original as possible. With his customarily punctilious and knowledge-

57 Sullivan, 'Ornamental bookbinding', 52–9.

58 Ibid. Sullivan's Spenser binding is illustrated and described in *Irish bookbinding, Dublin and provincial, De Búrca Rare book catalogue 31* (Winter 1993), no. 55.

8 Upper cover of binding finished by Sir Edward Sullivan, in tan morocco, gold-tooled, with black and crimson onlays, signed with 'ES' monogram, *c.*1900, on *La Tabaccheide, Ditirambo di Girolamo Baruffaldi* (Ferrara, 1714), photograph courtesy of Eamonn de Búrca, antiquarian bookseller, Dublin)

able approach, he analyzed the system, art and technique of this ninth-century miracle of survival, whose 339 pages had been mutilated (ironically) by an eighteenth-century bookbinder.

ACKNOWLEDGMENTS

Robert Ansell, Charles Benson (Rare Books Librarian, Trinity College, Dublin), Dr Maurice Craig, Alan Crawford, Julia Cummins (Librarian, Farmleigh, County Dublin), Eamonn de Búrca, Aidan Heavey, Alan Jutzi (Chief Curator, Rare Books, Huntington Library), the late Vincent Kinane (The Old Library, Trinity College, Dublin), Bryan D. Maggs, Philippa Marks, Dr Joseph McDonnell, Philippa Marks (Keeper of Bookbindings, British Library), Trish Quigley (Librarian, Milltown Park, Dublin).

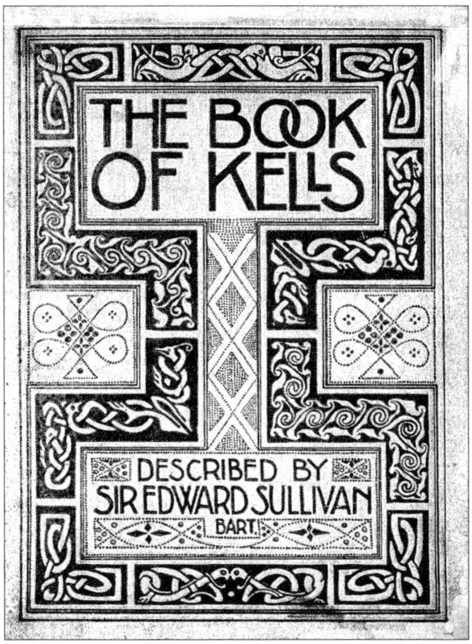

9 Cloth upper of *The Book of Kells*, described by Sir Edward Sullivan, Bart. (London, 1914)

Appendix

*Books published by Michael Adams at Four Courts Press, 1977–2005**

Compiled by Brad Morrow

Adams, Michael, *The hard life: religion for young adults* (1977)
Adams, Michael, *Single-minded: a tract on chastity, mainly in the context of celibacy* (1979)
Agnew, Jean, *Belfast merchant families in the seventeenth century* (1996)
Aldous, Richard, *Macmillan, Eisenhower and the Cold War* (2005)
Allen, Nicholas & Aaron Kelly (eds), *The cities of Belfast* (2003)
Allen, Nicholas, *George Russell (AE) and the new Ireland, 1905–30* (2003)
Alviar, J. José, *Klesis: the theology of the Christian vocation according to Origen* (1994)
Andrews, J.H., *A paper landscape: the Ordnance Survey in nineteenth-century Ireland* (2002)
Anon., *After communion: a book of devotions based on the psalms, the liturgy, spiritual writings & traditional prayers* (1988)
Augusteijn, Joost (ed.), *Irish history: a research yearbook 1* (2002)
Augusteijn, Joost & Mary Ann Lyons (eds), *Irish history: a research yearbook 2* (2003)
Augusteijn, Joost (ed.), *Ireland in the 1930s: new perspectives* (1999)
Baranski, Zygmunt G. (ed.), *Pasolini old and new: surveys and studies* (1999)
Barnard, Toby, *A guide to sources for the history of material culture in Ireland, 1500–2000* (2005)
Barnard, Toby, *Irish Protestant ascents and descents, 1641–1770* (2004)
Bartlett, Thomas (eds), *Revolutionary Dublin, 1795–1801: the letters of Francis Higgins to Dublin Castle* (2004)
Bartlett, Thomas [et al.] (ed.), *1798: a bicentenary perspective* (2003)

* The listing excludes a small number of publications for which Four Courts Press acted as a distributor rather than a publisher.

Baur, Benedict, *Frequent confession: its place in the spiritual life* (1984)
Beatty, John D. (ed.), *Protestant women's narratives of the Irish rebellion of 1798* (2001)
Belanger, Jacqueline (ed.), *The Irish novel in the nineteenth century: facts and fictions* (2005)
Belda, M. [et al.], *Holiness and the world: studies in the teachings of Blessed Josemaría Escrivá* (1997)
Bhreathnach, Edel (ed.), *The kingship and landscape of Tara* (2005)
Blackstock, Allan, *An Ascendancy army: the Irish Yeomanry, 1796–1834* (1998)
Boardman, Steve & Alasdair Ross (eds), *The exercise of power in medieval Scotland,* c.*1200–1500* (2003)
Boffey, Julia & John Scattergood (eds), *Texts and their contexts: papers from the Early Book Society* (1997)
Boklund-Lagopoulou, Karin, *'I have a yong suster': popular song and the Middle English lyric* (2001)
Bonsall, Penny, *The Irish RMs: the resident magistrates in the British administration of Ireland* (1997)
Bourke, Cormac (ed.), *Studies in the cult of Saint Columba* (1997)
Boyce, D. George & Roger Swift (eds), *Problems and perspectives in Irish history since 1800: essays in honour of Patrick Buckland* (2004)
Boyd, Gary A., *Dublin, 1745–1922: hospitals, spectacle and vice* (2005)
Boydell, Barra (ed.), *Music at Christ Church before 1800: documents and selected anthems* (1999)
Brady, Joseph & Anngret Simms (eds), *Dublin through space and time,* c.*900–1900* (2001)
Brand, Paul, Kevin Costello & W.N. Osborough (eds), *Adventures of the law: proceedings of the Sixteenth British Legal History Conference* (2005)
Brannigan, John, *Brendan Behan: cultural nationalism and the revisionist writer* (2002)
Breathnach, Ciara, *The Congested District Board of Ireland, 1891–1923: poverty and development in the West of Ireland* (2005)
Breen, Aidan (ed. and trans.), *Ailerani interpretatio mystica et moralis progenitorum Domini Iesu Christi* (1995)
Breen, Colin, *The Gaelic lordship of the O'Sullivan Beare: a landscape cultural history* (2005)
Breeze, Andrew, *Medieval Welsh literature* (1997)
Brennan, Michael, *The war in Clare, 1911–1921: personal memoirs of the Irish war of independence* (1980)
Brett, Edward M., *The British Auxiliary Legion in the first Carlist War, 1835–1838: a forgotten army* (2005)
Bright, Kevin, *The Royal Dublin Society, 1815–45* (2004)
Bristow, Peter E., *The moral dignity of man: Catholic moral teaching on family and medical ethics* (1993)
Brown, Michael & Stephen H. Harrison (eds), *The medieval world and the modern mind* (2000)
Brown, Michael, Charles Ivar McGrath & Thomas Power (eds), *Converts and conversion in Ireland, 1650–1850* (2005)

Brown, Michael, *Francis Hutcheson in Dublin, 1719–1730: the crucible of his thought* (2002)
Budnik, Mary Ann, *You can become a saint!* (1990)
Buehrle Marie C., *Cardinal Merry del Val Rafael* (1980)
Burke, Cormac, *Authority and freedom in the church* (1988)
Burke, Cormac, *Covenanted happiness: love and commitment in marriage* (1990)
Butler, David J., *South Tipperary, 1570–1841: religion, land and rivalry* (2005)
Byrne, Francis John, *Irish kings and high-kings* (2001)
Calder, George (ed.), *Auraicept na n-Éces: the scholars' primer* (1995)
Canals, Salvatore, *Jesus as friend* (1979)
Candon, Geraldine, *Headford, Co. Galway, 1775–1901* (2003)
Carey, John, *King of mysteries: early Irish religious writings* (1998)
Carey, John, Máire Herbert & Pádraig Ó Riain (eds), *Saints and scholars: studies in Irish hagiography* (2001)
Carey, Vincent & Ute Lotz-Heumann (eds), *Taking sides?: colonial and confessional* mentalités *in early modern Ireland* (2003)
Carey, Vincent, *Surviving the Tudors: the 'wizard' earl of Kildare and English rule in Ireland, 1537–1586* (2002)
Carroll, Ray, *Islands of the mind* (1989)
Carroll, Thomas K., *Wisdom and wasteland: the sermons of Jeremy Taylor today* (2001)
Carroll-Burke, Patrick, *Colonial discipline: the making of the Irish convict system* (2000)
Casciaro, J.M., *Jesus and politics* (1983)
Casciaro, J.M. & J.M. Monforte, *God, the world and man in the message of the Bible* (1996)
Castillo, Gerardo, *Teenagers and their problems* (1986)
Cavanna, Henry (ed.), *Governance, globalization and the European Union: which Europe for tomorrow?* (2002)
Chambers, Liam, *Michael Moore,* c.*1639–1726: provost of Trinity, rector of Paris* (2005)
Chambers, Liam, *Rebellion in Kildare, 1790–1803* (1998)
Chiu, Hungdaw (ed.), *The international law of the sea: supplement 1* (1997)
Clancy, Joseph P. (ed.), *Medieval Welsh poems* (2003)
Clare, Liam, *Enclosing the commons: Dalkey, the Sugar Loaves and Bray, 1820–1870* (2004)
Clark, Mary, Yvonne Desmond & Nodlaig P. Hardiman (eds), *Sir John T. Gilbert, 1829–1898: historian, archivist and librarian* (1999)
Clarke, Aidan, *The Old English in Ireland, 1625–42* (2000)
Clarke, Howard B., Máire Ní Mhaonaigh & Raghnall Ó Floinn (eds), *Ireland and Scandinavia in the early Viking age* (1998)
Clarke, Joe, *Christopher Dillon Bellew and his Galway estates, 1763–1826* (2003)
Clarkson, L.A., *A university in troubled times: Queen's Belfast, 1945–2000* (2004)
Clarkson, L.A. [et al.] (eds), *Mapping the Great Irish Famine: a survey of the famine decades* (1999)
Cliff, Brian & Éibhear Walshe (eds), *Representing the Troubles: texts and images, 1970–2000* (2004)

Cogan, Anthony, *The diocese of Meath: ancient and modern* (1992)
Cohen, Marilyn, *Linen, family, and community in Tullylish, County Down, 1690–1914* (1997)
Coldrick, Bryn, *Rossin, Co. Meath, 1724–1901* (2002)
Coll, Niall, *Christ in eternity and time: modern Anglican perspective* (2001)
Collins, Kevin, *Catholic churchmen and the Celtic revival in Ireland, 1848–1916* (2002)
Collins, Sinéad, *Balrothery Poor Law Union, County Dublin, 1839–1851* (2005)
Connell, Carmel, *Glasnevin Cemetery, Dublin, 1832–1900* (2004)
Connell, Desmond, *Christ our life: pastoral letters, 1988–95* (1995)
Connell, Desmond, *Essays in metaphysics* (1996)
Connell, Paul, Denis A. Cronin & Brian Ó Dalaigh (eds), *Irish townlands: studies in local history* (1998)
Connell, Paul, *The Diocese of Meath under Bishop John Cantwell 1830–66* (2004)
Connell, Peter, *The land and people of County Meath, 1750–1850* (2004)
Connolly, Charles (ed.), *New life* (1988)
Connolly, Charles (ed.), *On being Catholics* (1983)
Connolly, Hugh, *Irish penitentials and their significance for the sacrament of penance today* (1995)
Connolly, Philomena (ed.), *Statute rolls of the Irish Parliament: Richard III–Henry VIII* (2002)
Connolly, Philomena, *Medieval record sources* (2002)
Connolly, Ruth & Ann Coughlan (eds), *New voices in Irish criticism 5* (2004)
Connolly, S.J. (ed.), *Kingdoms United?: Great Britain and Ireland since 1500: integration and diversity* (1999)
Connolly, S.J. (ed.), *Political ideas in eighteenth-century Ireland* (2000)
Connolly, S.J., *Priests and people in pre-famine Ireland, 1780–1845* (2001)
Connolly, Seán & Diarmaid Scully, *Bede on Tobit and on the Canticle of Habakkuk* (1997)
Conrad O'Briain, Helen, Anne Marie D'Arcy & John Scattergood (eds), *Text and gloss: studies in insular learning and literature presented to Joseph Donovan Pheifer* (1999)
Conwell, John Joseph, *A Galway landlord during the Great Famine: Ulick John de Burgh, first marquis of Clanricarde* (2003)
Cooke, Colman, *Mary Charles Walker: the nun of Calabar* (1980)
Cooney, Helen (ed.), *Nation, court and culture: new essays on fifteenth-century English poetry* (2000)
Corcoran, Mary P. & Mark O'Brien (eds), *Political censorship and the democratic state: the Irish broadcasting ban* (2005)
Corish, Patrick J. & Benignus Millett (eds), *The Irish martyrs* (2005)
Cosgrave, Dillon, *North Dublin: city and county* (1977)
Cousins, Mel, *The birth of social welfare in Ireland, 1922–52* (2003)
Cox, Gareth & Axel Klein (eds), *Irish music in the twentieth century* (2003)
Crane, Ralph J. & Jennifer Livett (eds), *Troubled pleasures: the fiction of J.G. Farrell* (1997)
Crane, Ralph J. (ed.), *J.G. Farrell: the critical grip* (1999)

Crawford, E. Margaret, *Counting the people: a survey of the Irish censuses, 1813–1911* (2002)
Crawford, John, *The Church of Ireland in Victorian Dublin* (2005)
Crawford, Jon G., *A Star Chamber Court in Ireland: the Court of Castle Chamber, 1571–1641* (2005)
Cronin, Denis A., Jim Gilligan & Karina Holton (eds), *Irish fairs and markets: studies in local history* (2001)
Cronin, Elizabeth, *Fr Michael Dungan's Blanchardstown, 1836–68* (2002)
Cronin, Michael & Cormac Ó Cuilleanáin (eds), *The languages of Ireland: strangers to ourselves* (2003)
Cronin, Mike, *Sport and nationalism in Ireland: Gaelic games, soccer and Irish identity since 1884* (1999)
Cronin, Mike, *The Blueshirts and Irish politics* (1997)
Crowe, Ian (ed.), *Edmund Burke: his life and legacy* (2000)
Culleton, Edward, *Celtic and early Christian Wexford, 400–1166: AD 400 to 1160* (1999)
Cunningham, Bernadette & Raymond Gillespie, *Stories from Gaelic Ireland: microhistories from the sixteenth-century Irish annals* (2003)
Cunningham, Bernadette, *The world of Geoffrey Keating: history, myth and religion in seventeenth-century Ireland* (2000)
Curtis, Ruth (ed.), *When hope and history rhyme: the NUI, Galway, Millennium lecture series* (2002)
Daly, Cahal B., *Law and morals* (1993)
Daly, Cahal B., *Moral philosophy in Britain: from Bradley to Wittgenstein* (1996)
Daly, Cahal B., *Tertullian the Puritan and his influence* (1993)
Daly, Edward & Kieran Devlin, *The clergy of the diocese of Derry: an index* (1997)
Daly, Edward, *Mister, are you a priest?: jottings by Bishop Edward Daly* (2000)
Daly, Karina, *Tom Walsh's opera: a history of the Wexford Festival, 1951–2004* (2004)
Daly, Kieran Anthony, *Catholic church music in Ireland, 1878–1903: 'Lyra Ecclesiastica' and the Cecilian reform movement* (1995)
D'Arcy, Anne Marie & Alan J. Fletcher (eds), *Studies in late medieval and early Renaissance texts in honour of John Scattergood: 'The key of all good remembrance'* (2005)
D'Arcy, Anne Marie, *Wisdom and the Grail: the image of the vessel in the* Queste del Saint Graal *and Malory's* Tale of Sankgreal (2000)
D'Arcy, Gordon, *Ireland's lost birds* (1999)
de Margerie, Bertrand, *Daily communion and frequent confession* (1992)
de Paor, Liam, *Ireland and early Europe: essays and occasional writings on art and culture* (1997)
de Paor, Liam, *Landscapes with figures: people, culture and art in Ireland and the modern world* (1998)
de Paor, Liam, *On the Easter proclamation and other declarations* (1997)
de Paor, Liam, *Saint Patrick's world: the Christian culture of Ireland's apostolic age* (1993)
del Valle, Francisca Javiera, *About the Holy Spirit* (1981)
Delehaye, Hippolyte, *The legends of the saints* (1998)

Devine, Patrick F. & Harry White (eds), *Irish Musical Studies 4: the Maynooth international musicological conference 1995: selected proceedings* (1996)
Devine, Patrick F. & Harry White (eds), *Irish Musical Studies 5: the Maynooth international musicological conference 1995: selected proceedings* (1996)
Dillane, Fionnuala & Ronan Kelly (eds), *New voices in Irish criticism 4* (2003)
Dillon, Myles, *Early Irish literature* (1994)
Dillon, Myles, *The cycles of the kings* (1994)
Doherty, Gillian M., *The Irish Ordnance Survey: history, culture and memory* (2004)
Doherty, Peter (ed.), *Post-abortion syndrome: its wide ramifications* (1995)
Doherty, Peter & Agneta Sutton (eds), *Man-made man: ethical and legal issues in genetics* (1997)
Doherty, Richard & David Truesdale, *Irish winners of the Victoria Cross* (2000)
Doherty, Richard, *Ireland's generals in the Second World War* (2005)
Doherty, Richard, *Irish men and women in the Second World War* (1999)
Doherty, Richard, *Irish volunteers in the Second World War* (2001)
Doherty, Richard, *The Williamite War in Ireland, 1688–1691* (1998)
Dolan, T.P. & Diarmaid Ó Muirithe, *The dialect of Forth and Bargy, Co. Wexford, Ireland* (1996)
Dooley, Terence, *Inniskeen, 1912–1918: the political conversion of Bernard O'Rourke* (2004)
Doorley, Michael, *Irish-American diaspora nationalism: the Friends of Irish Freedom, 1916–35* (2005)
Dorgan, Theo (ed.), *Irish poetry since Kavanagh* (1996)
Dornan, Brian, *Mayo's lost islands: the Inishkeas* (2000)
Douglas, Aileen, Patrick Kelly & Ian Campbell Ross (eds), *Locating Swift: essays from Dublin on the 250th anniversary of the death of Jonathan Swift, 1667–1745* (1998)
Dryburgh, Paul & Brendan Smith (eds), *Handbook and select calendar of sources for medieval Ireland in the National Archives of the United Kingdom* (2005)
Dudley, Rowena, *The Irish lottery, 1780–1801* (2004)
Duffy, Patrick J., David Edwards & Elizabeth FitzPatrick (eds), *Gaelic Ireland, c.1250–c.1650: land, lordship, and settlement* (2001)
Duffy, Seán (ed.), *Medieval Dublin I–VI: proceedings of the Friends of Medieval Dublin Symposium 1999–2004* (6 vols) (2000)
Dungan, Myles, *They shall grow not old: Irish soldiers and the Great War* (1997)
Dunton, John, *Teague land: or, A merry ramble to the wilde Irish* (1980)
Dunton, John (ed. and intro. by Andrew Carpenter), *Teague land: or, A merry ramble to the wild Irish* (2003)
Dunton, John (ed. and intro. by Andrew Carpenter), *The Dublin scuffle* (2000)
Edel, Doris (ed.), *Cultural identity and cultural integration: Ireland and Europe in the early Middle Ages* (1995)
Edel, Doris, *The Celtic West and Europe: studies in Celtic literature and the early Irish Church* (2001)
Edwards, David (ed.), *Regions and rulers in Ireland, 1100–1650: essays in honour of Kenneth Nicholls* (2004)

Edwards, David, *The Ormond lordship in County Kilkenny, 1515–1642: the rise and fall of Butler feudal power* (2003)
Edwards, Francis, *Plots and plotters in the reign of Elizabeth I* (2002)
Edwards, Francis, *The succession, bye and main plots of 1601–1603* (2005)
Enright, Michael J., *Lady with a mead cup: ritual, prophecy and lordship in the European warband from La Tene to the Viking age* (1996)
Escrivá, Josémaría, *Conversations with Mgr Escrivá de Balaguer* (1980)
Escrivá, Josémaría, *An chor*ó*in mhuire naofa* (1979)
Escrivá, Josémaría, *Christ is passing by* (1982)
Escrivá, Josémaría, *Friends of God* (1981)
Escriva, Josémaría, *Holy rosary* (1979)
Escrivá, Josémaría, *The way* (1977)
Escrivá, Josémaría, *The way of the cross* (1983)
Escrivá, Josémaría, *Bealach* (1980)
Fagan, Patrick, *Catholics in a Protestant country: the Papist constituency in eighteenth-century Dublin* (1998)
Fagan, Patrick, *An Irish Bishop in penal times: the chequered career of Sylvester Lloyd OFM, 1680–1747* (1993)
Fagan, Patrick, *Divided loyalties: the question of an oath for Irish Catholics in the eighteenth century* (1997)
Fagan, Patrick, *Ireland in the Stuart papers: correspondence and documents of Irish interest from the Stuart papers in the Royal Archives, Windsor Castle* (1995)
Fagan, Patrick, *The Diocese of Meath in the eighteenth century* (2001)
Fallon, Brian, *Imogen Stuart: sculptor* (2002)
Fallon, Rosaleen, *A County Roscommon wedding, 1892: the marriage of John Hughes and Mary Gavin* (2004)
Fauske, Christopher J. (ed.), *Archbishop William King and the Anglican Irish context, 1688–1729* (2004)
Feheney, J.M., *Catholic education in Trinidad in the nineteenth century* (2001)
Fennell, Desmond, *Revision of Irish Nationalism* (1989)
Finan, Thomas & Vincent Twomey (eds), *Patristic Christology* (1998)
Finan, Thomas & Vincent Twomey (eds), *Scriptural interpretation in the Fathers: letter and spirit* (1995)
Finan, Thomas & Vincent Twomey (eds), *The relationship between neoplatonism and Christianity* (1992)
Fischer, Joachim & John Dillon (eds), *The correspondence of Myles Dillon, 1922–1925: Irish-German relations and Celtic studies* (1999)
FitzPatrick, Elizabeth & Raymond Gillespie (eds), *The parish in medieval and early modern Ireland: community, territory and building* (2005)
FitzSimon, E.A. Taylor & James H. Murphy (eds), *The Irish revival reappraised* (2003)
Fleming, John, *Gille of Limerick (*c.*1070–1145): architect of a medieval church* (2001)
Fleming, John, *St John's Cathedral, Limerick* (1987)
Fletcher, Alan J. & Raymond Gillespie (eds), *Irish preaching, 700–1700* (2001)
Fletcher, Alan J., *Preaching, politics and poetry in late-medieval England* (1998)

Flynn, Thomas S., *The Irish Dominicans, 1536–1641* (1993)
Foley, Tadhg & Seán Ryder (eds), *Ideology and Ireland in the nineteenth century* (1998)
Foley, Tadhg (ed.), *From Queen's College to National University: essays on the academic history of QCG/UCG/NUI, Galway* (2000)
Ford, Alan, *The Protestant Reformation in Ireland, 1590–1641* (1997)
Forrestal, Alison, *Catholic synods in Ireland, 1600–1690* (1998)
Foster, Sally W. (ed.), *The St Andrews Sarcophagus: a Pictish masterpiece and its international connections* (1998)
Friel, Mary, *Dancing as a social pastime in the south-east of Ireland, 1800–1897* (2004)
Froude-Durix, Carole (ed. Helen Lanigan Wood), *The life and art of Kathleen Bridle* (1998)
Fry, Susan Leigh, *Burial in medieval Ireland, 900–1500: a review of the written sources* (1999)
Fuentes, Antonio, *A guide to the Bible* (1987)
Fulton, Helen (ed.), *Medieval Celtic literature and society* (2005)
Gallagher, Ronan, *Violence and nationalist politics in Derry City, 1920–1923* (2003)
Geary, Laurence M. (ed.), *Rebellion and remembrance in modern Ireland* (2001)
Geiger, Till & Michael Kennedy (eds), *Ireland, Europe and the Marshall Plan* (2004)
Gibbons, Stephen Randolph, *Captain Rock, night errant: the threatening letters of pre-Famine Ireland, 1801–1845* (2004)
Gillen, Gerard & Andrew Johnstone (eds), *An historical anthology of Irish church music* (1999)
Gillespie, Raymond (ed.), *The proctor's accounts of Peter Lewis, 1564–1565* (1996)
Gillespie, Raymond (ed.), *The remaking of modern Ireland, 1750–1950: Beckett Prize essays in Irish history, 1999–2000* (2004)
Gillespie, Raymond (ed.), *The vestry records of the parish of St John the Evangelist, Dublin, 1595–1658* (2002)
Gillespie, Raymond (ed.), *The vestry records of the parishes of St Catherine and St James, Dublin, 1657–1692* (2004)
Gillespie, Raymond (ed.), *The first chapter act book of Chirst* [sic] *Church Cathedral, Dublin, 1574–1634* (1997)
Gillespie, Raymond & W.G. Neely (eds), *The laity and the Church of Ireland, 1000–2000: all sorts and conditions* (2002)
Gillis, Alan A. & Aaron Kelly (eds), *Critical Ireland: new essays in literature and culture* (2001)
Gilson, Etienne, *The unity of philosophical experience* (1982)
Girvin, Brian & Geoffrey Roberts (eds), *Ireland and the Second World War: politics, society and remembrance* (2000)
Gladitz, Charles, *Horse breeding in the medieval world* (1997)
Goodbody, Rob, *Sir Charles Domvile and his Shankill estate, County Dublin, 1857–1871* (2003)
Goodison, Richard, *Visions of the King* (1984)
Gormally, Luke (ed.), *Moral truth and moral tradition: essays in honour of Peter Geach and Elizabeth Anscombe* (1994)

Gougaud, Dom Louis (with intro. by Jean-Michel Picard), *Christianity in Celtic lands: a history of the churches of the Celts, their origin, their development, influence, and mutual relations* (1992)

Gray, Peter (ed.), *Victoria's Ireland? Irishness and Britishness, 1837–1901* (2004)

Greer, Desmond & James W. Nicolson, *The Factory Acts in Ireland, 1802–1914* (2003)

Greer, Desmond & Norma Dawson (eds), *Mysteries and solutions in Irish legal history* (2001)

Greshake, Gisbert, *The meaning of Christian priesthood* (1988)

Gribben, Crawford, *The Puritan millennium: literature and theology, 1550–1682* (2000)

Griffin, Brian, *Sources for the study of crime in Ireland, 1801–1921* (2005)

Grimes, Seamus (ed.), *Ireland in 1804* (1980)

Gurrin, Brian, *Pre-census sources for Irish demography* (2002)

Gwynn, Aubrey (edited by Gerard O'Brien), *The Irish church in the eleventh and twelfth centuries* (1993)

Hagger, Mark S., *The fortunes of a Norman family: the de Verduns in England, Ireland and Wales, 1066–1316* (2001)

Haines, Robin, *Charles Trevelyan and the Great Irish Famine* (2004)

Hall, Dianne, *Women and the church in medieval Ireland,* c.*1140–1540* (2002)

Hall, Donal, *World War I and nationalist politics in County Louth, 1914–1920* (2005)

Hall, Eamonn G. & Daire Hogan (eds), *The Law Society of Ireland, 1851–2001: portrait of a profession* (2002)

Halpin, Andrew (ed.), *The port of medieval Dublin: archaeological excavations at the Civic Offices, Winetavern Street, Dublin, 1993* (2000)

Hamilton, Mark, *Family matters: a guide to Christian family living* (1992)

Hanley, Brian (ed.), *A guide to Irish military heritage* (2004)

Hanley, Brian, *The IRA, 1926–1936* (2002)

Hanratty, Gerald, *Light from Paris: Cardinal Lustiger on faith and contemporary culture* (1995)

Hanratty, Gerald, *Philosophers of the enlightenment: Locke, Hume and Berkeley revisited* (1995)

Hanratty, Gerald, *Studies in Gnosticism and in the philosophy of religion* (1997)

Harbison, Peter, *Beranger's antique buildings of Ireland* (1998)

Harrison, Richard S., *A biographical dictionary of Irish Quakers* (1997)

Hart, A.R., *A history of the King's Serjeants at Law in Dublin: honour rather than advantage?* (2000)

Harvey, Karen J., *The Bellews of Mount Bellew: a Catholic gentry family in eighteenth-century Ireland* (1998)

Hawkes, Jane, *The Sandbach crosses: sign and significance in Anglo-Saxon sculpture* (2002)

Haywood, Eric G. (ed.), *Dante metamorphoses: episodes in a literary afterlife* (2003)

Heaney, Henry (ed.), *A Scottish Whig in Ireland, 1835–1838: the Irish journals of Robert Graham of Redgorton* (1999)

Helferty, Seamus & Raymond Refaussé (eds), *Directory of Irish archives* (4th ed., 2003)

Herbert, Máire & Kevin Murray (eds), *Retrospect and prospect in celtic studies: proceedings of the 11th International Congress of Celtic Studies* (2003)
Herbert, Máire, *Iona, Kells, and Derry: the history and hagiography of the monastic* familia *of Columba* (1996)
Herlihy, Jim, *The Royal Irish Constabulary: a short history and genealogical guide: with a select list of medal awards and casualties* (1997)
Herlihy, Jim, *Royal Irish Constabulary officers: a biographical dictionary and genealogical guide, 1816–1922* (2005)
Herlihy, Jim, *The Dublin Metropolitan Police: a complete alphabetical list of officers and men, 1836–1925* (2001)
Herlihy, Jim, *The Dublin Metropolitan Police: a short history and genealogical guide with notes on medal awards and casualties* (2001)
Herlihy, Jim, *The Royal Irish Constabulary: a complete alphabetical list of officers and men, 1816–1922* (1999)
Herlihy, Kevin (ed.), *Propagating the word of Irish dissent, 1650–1800* (1999)
Herlihy, Kevin (ed.), *The Irish dissenting tradition, 1650–1750* (1995)
Herlihy, Kevin (ed.), *The politics of Irish dissent 1650–1800* (1997)
Herlihy, Kevin (ed.), *The religion of Irish dissent, 1650–1800* (1996)
Hickey, Éanna, *Irish law and lawyers in modern folk tradition* (1999)
Hill, John & Kevin Rockett (eds), *Film history and national cinema* (2005)
Hill, Judith, *Irish public sculpture: a history* (1998)
Hirst, Catherine, *Religion, politics and violence in nineteenth-century Belfast: the Pound and Sandy Row* (2002)
Hogan, Edmund, *Onomasticon Goedelicum locorum et tribuum Hiberniae et Scotiae: an index,with identifications, to the Gaelic names of places and tribes* (1993)
Holton, Karina, Liam Clare & Brian Ó Dálaigh (eds), *Irish villages: studies in local history* (2003)
Holzherr, George (ed.), *The rule of Saint Benedict: a guide to Christian living: the full text of the rule in Latin and English* (1994)
Hooper, Glenn & Leon Litvack (eds), *Ireland in the nineteenth century: regional identity* (2000)
Horgan, John, *Broadcasting and public life: RTÉ news and current affairs, 1926–1997* (2004)
Hourihane, Colum (ed.), *Irish art historical studies in honour of Peter Harbison* (2004)
Howlett, David R., *The English origins of Old French literature* (1996)
Howlett, David R., *British books in biblical style* (1997)
Howlett, David R. (ed. and trans.), *Liber epistolarum Sancti Patricii episcopi: The book of letters of Saint Patrick the Bishop* (1994)
Howlett, David R., *The Celtic Latin tradition of biblical style* (1995)
Howlett, David R., *Caledonian craftmanship: the Scottish Latin tradition* (2000)
Howlett, David R., *Cambro-Latin compositions: their competence and craftsmanship* (1998)
Howlett, David R., *Insular inscriptions* (2005)
Howlett, David R., *Sealed from within: self-authenticating insular charters* (1999)
Huber, Georges, *My angel will go before you* (1983)

Hughes, Kathleen & Ann Hamlin, *The modern traveller to the early Irish Church* (1997)
Illanes, José Luis, *On the theology of work: aspects of the teaching of the founder of Opus Dei* (1982)
Isaacs, David, *Character building: a guide for parents and teachers* (1984)
Jackson, Kenneth Hurlstone, *Language and history in early Britain: chronological survey of the Brittonic languages, 1st to 12th century AD* (1994)
James, Francis G. & Miriam G. Hill (eds), *Joy to the world: two thousand years of Christmas* (2000)
Jaski, Bart, *Early Irish kingship and succession* (2000)
Jefferies, Henry A. & Ciarán Devlin (eds), *History of the Diocese of Derry from earliest times* (2000)
Jefferies, Henry A. (ed.), *History of the Diocese of Clogher* (2005)
Jefferies, Henry A. (with contributions by Gerard O'Brien and Eilis C. Stack), *Cork: historical perspectives* (2004)
Jefferies, Henry A., *Priests and prelates of Armagh in the age of reformations, 1518–1558* (1997)
Jungmann, Joseph A., *The Mass of the Roman rite: its origins and development (Missarum Sollemnia)* (1986)
Kabdebo, Thomas, *Ireland and Hungary: a study in parallels: with an Arthur Griffith bibliography* (2001)
Keenan, Celia & Mary Shine Thompson (eds), *Studies in children's literature, 1500–2000* (2004)
Keenan, Celia & Mary Shine Thompson (eds), *Treasure Islands: studies in children's literature* (2005)
Kelly, Bernard, *Introduction to moral theology: fundamental concepts in their Christian perspective* (1983)
Kelly, James & Dáire Keogh (eds), *History of the Catholic Diocese of Dublin* (1999)
Kelly, James, *Gallows speeches from eighteenth-century Ireland* (2001)
Kelly, James, *Henry Flood: patriots and politics in eighteenth-century Ireland* (1998)
Kelly, James, *Sir Edward Newenham, MP, 1734–1814: defender of the Protestant constitution* (2004)
Kelly, James, *The Liberty and Ormond Boys: factional riot in eighteenth-century Dublin* (2005)
Kelly, Liam, *Kiltubrid, County Leitrim: snapshots of a rural parish in the 1890s* (2005)
Kelly, William & John R. Young (eds), *Ulster and Scotland, 1600–2000: history, language and identity* (2004)
Kelly, William (ed.), *The sieges of Derry* (2001)
Kennedy, Michael & Joseph Morrison Skelly (eds), *Irish foreign policy, 1919–66: from independence to internationalism* (1999)
Kennedy, S.B., *David Crone: paintings, 1963–1999* (1999)
Kennedy, S.B., *T. P. Flanagan* (1995)
Kennedy, Angela (ed.), *Swimming against the tide: feminist dissent on the issue of abortion* (1997)

Kennedy, Edward, *The land movement in Tullaroan, County Kilkenny, 1879–1891* (2004)
Kennedy-Andrews, Elmer, *Fiction and the Northern Ireland Troubles since 1969: (de)constructing the north* (2003)
Kenney, James F., *The sources for the early history of Ireland: Ecclesiastical* (1993)
Kenny Colum, *Kilmainham: the history of a settlement older than Dublin with an account of its founder Saint Maignenn and of its holy well* (1995)
Kenny, Colum, *King's Inns and the battle of the books, 1972: cultural controversy at a Dublin library* (2002)
Kenyon, John R. & Kieran O'Conor (eds), *The medieval castle in Ireland and Wales: essays in honour of Jeremy Knight* (2003)
Keogh, Dáire & Nicholas Furlong (eds), *The women of 1798* (1998)
Keogh, Dáire & Nicholas Furlong (eds), *The mighty wave: the 1798 rebellion in Wexford* (1996)
Keogh, Dáire & Kevin Whelan (eds), *Acts of Union: the causes, contexts, and consequences of the Act of Union* (2001)
Keogh, Dáire, *Edmund Rice, 1762–1844* (1996)
Keogh, Dáire, '*The French disease*': *the Catholic Church and Irish radicalism, 1790–1800* (1993)
Keynes, Simon & Alfred P. Smyth (eds), *Anglo Saxons: studies presented to Cyril Roy Hart* (2005)
Kiberd, Damien (ed.), *Media in Ireland: ethics in broadcasting* (2002)
Kiberd, Damien (ed.), *Media in Ireland: the search for diversity* (1997)
Kiberd, Damien (ed.), *Media in Ireland: the search for ethical journalism* (1999)
Killeen, Jarlath, *Gothic Ireland: horror and the Irish Anglican imagination in the long eighteenth century* (2005)
Kinane, Vincent & Anne Walsh (eds), *Essays on the history of Trinity College Library, Dublin* (2001)
Kingston, Simon, *Ulster and the isles in the fifteenth century: the lordship of the Clan Domhnaill of Antrim* (2004)
Kotsonouris, Mary, *The winding up of the Dáil courts, 1922–1925: an obvious duty* (2004)
Lacey, Brian, *Colum Cille and the Columban tradition* (1997)
Laird, Heather, *Subversive law in Ireland, 1879–1920: from 'unwritten law' to the Dáil courts* (2005)
Langan-Egan, Maureen, *Galway women in the nineteenth century* (1999)
Larkin, Emmet, *The historical dimensions of Irish Catholicism* (1997)
Larkin, Emmet, *The Roman Catholic Church and the emergence of the modern Irish political system, 1874–1878* (1996)
Leahy, Anne & Yo Tomita (eds), *Bach studies from Dublin: Irish Musical Studies: 8* (2004)
Lee, Gerard A., *Leper hospitals in medieval Ireland: with a short account of the Military and Hospitaller Order of St Lazarus of Jerusalem* (1996)
Legg, Marie-Louise, *Newspapers and nationalism: the Irish provincial Press, 1850–1892* (1998)

Legg, Marie-Louise, *The diary of Nicholas Peacock, 1740–51: the worlds of a county Limerick farmer and agent* (2005)
Lennon, Colm, *An Irish prisoner of conscience of the Tudor era: Archbishop Richard Creagh of Armagh, 1523–1586* (2000)
Lennon, Colm, *Richard Stanihurst the Dubliner* (1980)
Lewis, Gifford, *Edith Somerville: a biography* (2005)
Liszka, Thomas R. & Lorna E.M.Walker (eds), *The North Sea world in the Middle Ages: studies in the cultural history of North-Western Europe* (2000)
Lucey, Donnacha Seán, *The Irish National League in Dingle, County Kerry, 1885–1892* (2003)
Luddy, Maria (ed.), *The Crimean journals of the Sisters of Mercy, 1854–56* (2004)
Lydon, James, *Law and Disorder in thirteenth-century Ireland: the Dublin Parliament of 1297* (1997)
Lydon, James, *The lordship of Ireland in the Middle Ages* (2003)
Lyons, Mary Ann, *Church and society in County Kildare,* c.*1470–1547* (2000)
Lysaght, Moira, *Fr Theobald Mathew OFM: the apostle of temperance* (1983)
Lysaght, Seán, *Robert Lloyd Praeger: the life of a naturalist* (1998)
Lyttleton, James & Tadhg O'Keeffe (eds), *The manor in medieval and early modern Ireland* (2003)
Macalister, R.A.S., *Corpus inscriptionum insularum Celticarum* (1996)
McAreavey, John, *The canon law of marriage and the family* (1997)
Macaulay, Ambrose, *William Crolly: Archbishop of Armagh, 1835–49* (1994)
Macaulay, Ambrose, *The Holy See, British policy and the plan of campaign in Ireland, 1885–93* (2002)
McAvera, Brian, *Marking the North: the work of Victor Sloan* (1990)
McAvera, Brian, *Art, politics and Ireland* (1990)
McBride, Ian, *The siege of Derry in Ulster Protestant mythology* (1997)
McBride, Lawrence W. (ed.), *Images, icons, and the Irish nationalist imagination* (1999)
McBride, Lawrence W. (ed.), *Reading Irish histories: texts, contexts, and memory in modern Ireland* (2003)
McCarroll, Joseph, *Marriage or divorce: the real issue* (1985)
McCarthy, Brian (ed.), *The Gospel according to St Mark* (1985)
McCarthy, Conor, *Modernisation, crisis and culture in Ireland, 1969–1992* (2000)
McCarthy, Daniel P. & Aidan Breen (eds), *The ante-Nicene Christian pasch: De ratione paschali: the paschal tract of Anatolius, bishop of Laodicea* (2002)
McCarthy, Michael (ed.), *Lord Charlemont and his circle: essays in honour of Michael Wynne* (2001)
McCarthy, Michael, *Classical and Gothic: studies in the history of art* (2005)
McCarthy, Muriel & Ann Simmons (eds), *The making of Marsh's Library: learning, politics and religion in Ireland, 1650–1750* (2004)
McCarthy, Muriel, *Marsh's library, Dublin: all graduates & gentlemen* (2003)
MacCarthy, Robert, *Ancient and modern: a short history of the Church of Ireland* (1995)
MacCarthy, Robert, *How shall they hear?: sermons and addresses* (2002)

MacCarthy, Robert, *Preached at St Patrick's, Dublin* (2004)
McCormack, Anthony, *The earldom of Desmond, 1463–1583: the decline and crisis of a feudal lordship* (2005)
McCormack, Bridget, *Perceptions of St Patrick in eighteenth-century Ireland* (2000)
MacCotter, Paul, *Colmán of Cloyne: a study* (2004)
McCracken, Donal P., *MacBride's brigade: Irish commandos in the Anglo-Boer War* (1999)
McEnery, M.J. & Raymond Refaussé (eds), *Christ Church deeds* (2001)
McEvoy, James & Michael Dunne (eds), *Thomas Aquinas: approaches to truth: the Aquinas lectures at Maynooth, 1996–2001* (2002)
McGee, Owen, *The IRB: the Irish Republican Brotherhood, from the Land League to Sinn Féin* (2005)
McGettigan, Darren, *Red Hugh O'Donnell and the Nine Years War* (2005)
McGinnity, Gerald, *Christmen: experience of priesthood today* (1985)
McGovern, Thomas J., *Priestly identity: a study in the theology of priesthood* (2002)
McGovern, Thomas J., *Priestly celibacy today* (1998)
McGrath, Charles Ivar, *The making of the eighteenth-century Irish constitution: government, parliament and the revenue, 1692–1714* (2000)
McGrath, Thomas, *Politics, interdenominational relations and education in the public ministry of Bishop James Doyle of Kildare and Leighlin, 1786–1834* (1999)
McGrath, Thomas, *Religious renewal and reform in the pastoral ministry of Bishop James Doyle of Kildare and Leighlin, 1786–1834* (1999)
McGrath, Thomas, *The pastoral and educational letters of Bishop James Doyle of Kildare and Leighlin, 1786–1834* (2003)
McGurk, John, *Sir Henry Docwra, 1564–1631: Derry's second founder* (2005)
McHugh, Damian, *Libel law: a journalist's handbook* (2001)
McIlvanney, Liam & Ray Ryan (eds), *Ireland and Scotland: culture and society, 1700–2000* (2005)
Macinnes, Allan & Jane Ohlmeyer (eds), *The Stuart kingdoms in the seventeenth century* (2002)
McKenny, Kevin, *The Laggan army in Ireland, 1640–1685: the landed interests, political ideologies and military campaigns of the north-west Ulster settlers* (2005)
McLoughlin, T.O., *Contesting Ireland: Irish voices against England in the eighteenth century* (1999)
McManus, Antonia, *The Irish hedge school and its books, 1695–1831* (2002)
McManus, Ruth, *Dublin, 1910–1940: shaping the city & suburbs* (2002)
McNally, Patrick, *Parties, patriots, and undertakers: parliamentary politics in early Hanoverian Ireland* (1997)
McNamara, Martin (ed.), *Apocalyptic and eschatological heritage: the Middle East and Celtic realms* (2003)
MacShamhráin, Ailbhe (ed.), *The island of St Patrick: church and ruling dynasties in Fingal and Meath, 400–1148* (2004)
Magennis, Eoin, *The Irish political system, 1740–1765: the golden age of the undertakers* (2000)

Magennis, Hugh, *Anglo-Saxon appetites: food and drink and their consumption in Old English and related literature* (1999)
Maginn, Christopher, *'Civilizing' Gaelic Leinster: the extension of Tudor rule in the O'Byrne and O'Toole lordships* (2005)
Malcomson, A.P.W., *Archbishop Charles Agar: churchmanship and politics in Ireland, 1760–1810* (2002)
Malcomson, A.P.W., *Nathaniel Clements: government and the governing elite in Ireland, 1725–75* (2005)
Martinez, Javier Garcia, *Mountain walks in Gran Canaria* (1998)
Mathews, P.J. (ed.), *New voices in Irish criticism* (2000)
Meek, Christine & Catherine Lawless (eds), *Pawns or players?: studies on medieval and early modern women* (2003)
Meek, Christine & Catherine Lawless (eds), *Victims or viragos?: studies on medieval and early modern women* (2005)
Meek, Christine & Katharine Simms (eds), *The fragility of her sex?: medieval Irish women in their European context* (1996)
Meek, Christine (ed.), *Women in late medieval and early modern Europe* (1999)
Meredith, Jane, *Around and about the Custom House* (1997)
Millar, Bonnie, *The siege of Jerusalem in its physical, literary, and historical contexts* (2000)
Mills, James (ed.), *Account roll of the Priory of the Holy Trinity, Dublin, 1337–1346* (1996)
Milne, Kenneth (ed.), *Christ Church Cathedral, Dublin: a history* (2000)
Milne, Kenneth, *The Irish charter schools, 1730–1830* (1997)
Moloney, Brian, *Italian novels of peasant crisis, 1930–50: bonfires in the night* (2005)
Morales, José, *Creation theology* (2001)
Moran, Gerard, *A radical priest in Mayo: Fr Patrick Lavelle: the rise and fall of an Irish nationalist, 1825–86* (1994)
Moran, Gerard, *Radical Irish Priests, 1660–1970* (1998)
Moran, Gerard (ed.), *Sending out Ireland's poor: assisted emigration to North America in the nineteenth century* (2004)
Morgan, Hiram (ed.), *Political ideology in Ireland, 1541–1641* (1999)
Morrissey, Thomas J., *Bishop Edward Thomas O'Dwyer of Limerick, 1842–1917* (2003)
Morrissey, Thomas J., *Thomas A. Finlay SJ, 1848–1940: educationalist, editor, social reformer* (2004)
Morrissey, Thomas J., *William J. Walsh, Archbishop of Dublin, 1841–1921* (2000)
Morrissey, Thomas J., *As one sent: Peter Kenney SJ, 1779–1841: his mission in Ireland and North America* (1996)
Mulcahy, Rosemarie, *Philip II of Spain: patron of the arts* (2004)
Mullally, Evelyn (ed. and trans.), *The deeds of the Normans in Ireland:* La geste des engleis en Yrlande (2002)
Munro, Margaret T., *The ragged saint: Benedict Joseph Labre* (1984)
Murphy, Anne M. & Eoin G. Cassidy (eds), *Neglected wells: spirituality and the arts* (1997)
Murphy, Daniel, *A history of Irish emigrant and missionary education* (2000)

Murphy, Daniel, *Christianity and modern European literature* (1997)
Murphy, David, *Ireland and the Crimean war* (2002)
Murphy, Gerard (ed.), *Early Irish lyrics: eighth to twelfth century* (1998)
Murphy, Ignatius, *The Diocese of Killaloe in the eighteenth century* (1991)
Murphy, Ignatius, *The Diocese of Killaloe, 1800–1850* (1992)
Murphy, Ignatius, *The diocese of Killaloe, 1850–1904* (1995)
Murphy, James H. (ed.), *Evangelicals and Catholics in nineteenth-century Ireland* (2005)
Murphy, James H., *Ireland: a social, cultural and literary history, 1791–1891* (2003)
Murphy, Sharon, *Maria Edgeworth and romance* (2004)
Murray, Patrick (ed.), *The deer's cry: a treasury of Irish religious verse* (1986)
Myers, Kevin, *Kevin Myers, from the* Irish Times *column 'An Irishman's Diary'* (2000)
Nagy, Joseph Falaky, *Conversing with angels and ancients: literary myths of medieval Ireland* (1997)
Nagy, Joseph Falaky (ed.), *Identifying the Celtic: CSANA yearbook 2* (2002)
Nagy, Joseph Falaky (ed.), *The individual in Celtic literatures: CSANA yearbook 1* (2001)
Nagy, Joseph Falaky & Leslie Ellen Jones (eds), *Heroic poets and poetic heroes in Celtic tradition: a festschrift for Patrick K. Ford* (2005)
Navarre Bible: The New Testament (standard edition: 12 volumes) (1986–1992)
Navarre Bible: The Old Testament (standard edition: 7 volumes) (1999–2005)
Navarre Bible: The New Testament (omnibus edition: 3 volumes) (2000–2005)
Navarre Bible: The New Testament (compact edition: 1 volume) (2001)
Neville, Cynthia J., *Native lordship in medieval Scotland: the earldoms of Strathearn and Lennox,* c.*1140–1365* (2005)
Newman, Jeremiah, *Balance in the church* (1980)
Newman, Jeremiah, *Ireland must choose* (1983)
Newman, Jeremiah, *Puppets of Utopia: can Irish democracy be taken for granted?* (1987)
Newman, Jeremiah, *Return to the sacred: a socio-religious analysis* (1986)
Newman, Jeremiah, *The post modern church* (1990)
Newman, Jeremiah, *The state of Ireland* (1977)
Newton, Michael, *A handbook of the Scottish Gaelic world* (2000)
Ní Bhroiméil, Úna, *Building Irish identity in America, 1870–1915: the Gaelic revival* (2003)
Ní Chatháin, Próinséas & Michael Richter (eds), *Ireland and Europe in the early Middle Ages: texts and transmission* (2001)
Ní Chuilleanáin, Eiléan (ed.), *The Wilde legacy* (2003)
Ní Mhurchadha, Maighread, *Fingal, 1603–60: contending neighbours in North Dublin* (2005)
Ní Riain, Íde M. (ed. and trans.), *Denis the Carthusian: spiritual writings: contemplation, meditation, prayer; the fountain of light and the paths of life; monastic profession; exhortation to novices* (2005)
Ní Uallacháin, Pádraigín, *A hidden Ulster: people, songs and traditions of Oriel* (2003)
Ní Mheara, Roisin, *In search of Irish saints: the Peregrinatio pro Christo* (1994)

Novick, Ben, *Conceiving revolution: Irish nationalist propaganda during the First World War* (2001)
Nugent, John D., *Nerves, worry and depression: the way out* (1991)
O'Brien, Gerard, *Irish governments and the guardianship of historical records, 1922–72* (2004)
O'Brien, John B., *Studies in Irish, British and Australian relations, 1916–63: trade diplomacy and politics* (2005)
O'Byrne, Emmett, *War, politics and the Irish of Leinster, 1156–1606* (2003)
Ó Cadhla, Stiofán, *The holy well tradition: the pattern of St Declan, Ardmore, County Waterford, 1800–2000* (2002)
O'Callaghan, Paul, *Fides Christi: the justification debate* (1997)
O'Callaghan, Paul, *The christological assimilation of the apocalypse: an essay on fundamental eschatology* (2004)
O'Callaghan, Paul [et al.] (eds), *Holiness in the World* (1998)
Ocáriz, F., L.F. Mateo Seco & J.A. Riestra, *The mystery of Jesus Christ: a christology and soteriology textbook* (1994)
Ó Ciardha, Éamonn, *Ireland and the Jacobite cause, 1685–1766: 'a fatal attachment'* (2002)
Ó Clabaigh, Colmán N., *The Franciscans in Ireland, 1400–1534: from reform to Reformation* (2002)
Ó Conbhuidhe, Colmcille (ed. Finbarr Donovan), *The Cistercian abbeys of Tipperary* (1999)
Ó Conbhuidhe, Colmcille (ed. Finbarr Donovan), *Studies in Irish Cistercian history* (1998)
O'Connell, Patricia, *The Irish College at Alcala de Henares, 1649–1785* (1997)
O'Connell, Patricia, *The Irish College at Lisbon, 1590–1834* (2001)
O'Connor, Thomas & Mary Ann Lyons (eds), *Irish migrants in Europe after Kinsale, 1602–1820* (2003)
O'Connor, Thomas (ed.), *The Irish in Europe, 1580–1815* (2001)
O'Connor, Thomas, *An Irish theologian in enlightenment France: Luke Joseph Hooke, 1714–96* (1995)
Ó Corráin, Donnchadh (ed.), *Irish antiquity: essays and studies presented to Professor M.J. O'Kelly* (1994)
Ó Corráin, Donnchadh (ed.), *James Hogan: revolutionary, historian and political scientist* (2001)
Ó Cróinín, Dáibhí, *The songs of Elizabeth Cronin, Irish traditional singer: the complete song collection* (2000)
Ó Cróinín, Dáibhí, *Early Irish history and chronology* (2001)
Ó Cuilleanáin, Cormac & Jennifer Petrie (eds), *Patterns in Dante: nine literary essays* (2005)
Ó Cuilleanáin, Donal, *Preparing for Confession* (1985)
O'Curry, Eugene, *Lectures on the manuscript materials of ancient Irish history* (1995)
Ó Dochartaigh, Pól, *Julius Pokorny, 1887–1970* (2003)
O'Donnell, Ian & Finbarr McAuley (eds), *Criminal justice history: themes and controversies from pre-independence Ireland* (2003)

O'Donnell, Manus (ed. and trans. by Brian Lacey), *The life of Colum Cille* (1998)
O'Driscoll, Mervyn, *Ireland, Germany and the Nazis: politics and diplomacy, 1919–1939* (2004)
O'Dwyer, Michael, *Julien Green: a critical study* (1997)
O'Dwyer, Peter, *Mary: a history of devotion in Ireland* (1988)
O'Farrell, Ciara, *Louis D'Alton and the Abbey Theatre* (2004)
O'Flynn, Críostóir, *A poet in Rome* (1992)
O'Grady, John, *The life and work of Sarah Purser* (1996)
Ohlmeyer, Jane H., *Civil war and restoration in the three Stuart Kingdoms: the career of Randal MacDonnell, Marquis of Antrim, 1609–1683* (2001)
O'Keeffe, Tadhg, *Romanesque Ireland: architecture and ideology in the twelfth century* (2003)
O'Leary, Cornelius & Patrick Maume, *Controversial issues in Anglo-Irish relations, 1910–1921* (2004)
O'Loughlin, Thomas (ed.), *Adomnán at Birr, AD 697: essays in commemoration of the law of the innocents* (2001)
Ó Maitiú, Séamas, *Dublin's suburban towns, 1834–1930: governing Clontarf, Drumcondra, Dalkey, Killiney, Kilmainham, Pembroke, Kingstown, Blackrock, Rathmines and Rathgar* (2003)
O'Meara, John J. (with paintings by Daniel De' Angeli), *The voyage of Saint Brendan* (1994)
O'Meara, John J., *Understanding Augustine* (1997)
Ó Muirithe, Diarmaid & Deirdre Nutall (eds), *The folklore of County Wexford* (1999)
Ó Muirithe, Diarmaid, *A dictionary of Anglo-Irish: words and phrases from Gaelic in the English of Ireland* (1996)
Ó Muirithe, Diarmaid, *A word in your ear* (1997)
Ó Muirithe, Diarmaid, *The words we use* (1996)
Ó Muirithe, Diarmaid, *The words we use: 3* (1999)
O'Regan, Philip, *Archbishop William King of Dublin (1650–1729) and the constitution in church and state* (2000)
Ó Riain, Flann, *Townlands of Leinster* (2000)
Orlandis, José, *A short history of the catholic Church* (1985)
O'Rourke, Brian, *The conscience of the race: sex and religion in Irish and French novels, 1941–1973* (1980)
Orpen, Goddard Henry, *Ireland under the Normans, 1169–1333* (2005)
Osborough, W.N., *Studies in Irish legal history* (1999)
O'Shaughnessy, Peter (ed.), *Rebellion in Wicklow: General Joseph Holt's personal account of 1798* (1998)
Ó Siochrú, Micheál (ed.), *Kingdoms in crisis: Ireland in the 1640s; essays in honour of Dónal Cregan* (2001)
Ó Siochrú, Mícheál, *Confederate Ireland, 1642–1649: a constitutional and political analysis* (1999)
O'Sullivan, Catherine, *Hospitality in medieval Ireland, 900–1500* (2004)
Pantos, Aliki & Sarah Semple (eds), *Assembly places and practices in medieval Europe* (2003)

Parkinson, Alan F., *Belfast's unholy war: the Troubles of the 1920s* (2004)
Parkinson, Alan F., *Ulster loyalism and the British media* (1998)
Patten, Eve, *Samuel Ferguson and the culture of nineteenth-century Ireland* (2004)
Patterson, Brad (ed.), *Ulster-New Zealand migration and cultural transfers* (2005)
Phelan, Diarmuid Rossa, *It's God we ought to crucify: validity and authority in law* (2000)
Phillips, Helen (ed.), *Robin Hood: medieval and post-medieval* (2005)
Picard, Jean-Michel (ed.), *Aquitaine and Ireland in the Middle Ages* (1995)
Picard, Jean-Michel (ed.), *Ireland and Northern France, AD 600–850* (1991)
Picard, Jean-Michel (trans.) (intro. by Yolande de Pontfarcy), *The vision of Tnugdal* (1989)
Picard, Jean-Michel (trans.) (intro. by Yolande de Pontfarcy), *Saint Patrick's purgatory: a twelfth century tale of a journey to the otherworld* (1985)
Pine, Richard, *2RN and the origins of Irish radio* (2002)
Pine, Richard, *Music and broadcasting in Ireland* (2005)
Plummer, Charles, *Vitae sanctorum Hiberniae* (1997)
Poppe, Erich & Bianca Ross (eds), *The legend of Mary of Egypt in medieval insular hagiography* (1996)
Potter, Simon J. (ed.), *Newspapers and empire in Ireland and Britain: reporting the British Empire,* c.*1857–1921* (2004)
Potterton, Michael, *Medieval Trim: history and archaeology* (2005)
Prunty, Jacinta, *Maps and map-making in local history* (2004)
Prunty, Jacinta, *Margaret Aylward, 1810–1889: lady of charity, sister of faith* (1999)
Purdue, Olwen, *The MacGeough Bonds of the Argory: an Ulster gentry family, 1880–1950* (2005)
Pye, Patrick, *The time gatherer: El Greco and the sacred theme* (1991)
Pyle, Hilary (ed.), *Deborah Brown: from painting to sculpture* (2005)
Raftery, Deirdre, *Women and learning in English writing, 1600–1900* (1997)
Refaussé, Raymond & Mary Clark, *A catalogue of the maps of the estates of the archbishops of Dublin, 1654–1850* (2000)
Refaussé, Raymond (ed.), *The registers of Christ Church Cathedral, Dublin* (1998)
Richter, Michael & Jean-Michael Picard (eds), *Ogma: essays in Celtic studies in honour of Próinséas Ní Chatháin* (2001)
Richter, Michael, *Ireland and her neighbours in the seventh century* (1999)
Richter, Michael, *Studies in medieval language and culture* (1995)
Richter, Michael, *The formation of the medieval West: studies in the oral culture of the barbarians* (1994)
Robinson-Hammerstein, Helga (ed.), *European universities in the age of Reformation and Counter Reformation* (1999)
Rockett, Kevin & John Hill (eds), *National cinema and beyond* (2004)
Rockett, Kevin, *Irish film censorship: a cultural journey from silent cinema to internet pornography* (2004)
Rodriguez, Pedro, *Personal prelatures and particular churches: a theological study of a new canonical institution* (1986)
Rohrbach, Peter, *Conversation with Christ* (1980)

Rooney, Brendan, *The life and work of Harry Jones Thaddeus, 1859–1929* (2003)
Roth, Andreas, *Mr Bewley in Berlin: aspects of the career of an Irish diplomat, 1933–1939* (2000)
Ryan, Arthur H. (ed. Ambrose Macaulay), *Mirroring Christ's splendour* (1984)
Ryan, John (ed.), *Féil-sgríbhinn Eóin Mhic Néill: essays and studies presented to Professor Eoin MacNeill* (1995)
Ryan, John, *Irish monasticism: origins and early development* (1993)
Ryan, Molly, *Over my shoulder: Kenya walkabout* (1987)
Sarkin, Jeremy & William Binchy (eds), *The administration of justice: current themes in comparative perspective* (2004)
The Sarum missal: missale ad usum insignis et praeclarae Ecclesiae Sarum: done into Engish and abridged (1989)
Savage, Robert J., *Ireland in the new century: politics, culture and identity* (2003)
Scattergood, John, *Reading the past: essays on medieval literature and society* (1996)
Scattergood, John, *The lost tradition: essays on Middle English alliterative poetry* (2000)
Scheffczyk, Leo, *On being a Christian: the Hans Kung debate* (1982)
Scott, Anne M., *Piers Plowman and the poor* (2004)
Selmer, Carl (ed.), *Navigatio Sancti Brendani abbatis: from early Latin manuscripts* (1989)
Sexton, Michael, *Marconi: the Irish connection* (2005)
Shannon, Pat, *Charles and me: notes in the margin* (2002)
Sheehan, Helena, *The continuing story of Irish television drama: tracking the tiger* (2004)
Silke, John J., *Kinsale: the Spanish intervention in Ireland at the end of the Elizabethan wars* (2000)
Simms, George Otto, *Angels & saints* (1988)
Simms, J.G., *Jacobite Ireland, 1685–91* (2000)
Sjoestedt, Marie-Louise (trans. Myles Dillon), *Gods and heroes of the Celts* (1994)
Sloan, Robert, *William Smith O'Brien and the Young Ireland Rebellion of 1848: the road to Ballingarry* (2000)
Smith, Louis P.F. & Sean Healy, *Farm organisations in Ireland: a century of progress* (1996)
Smyth, Alfred P. (ed.), *Seanchas: studies in early and medieval Irish archaeology, history and literature in honour of Francis J. Byrne* (2000)
Smyth, Alfred P., *Faith, famine and fatherland in the Irish midlands: perceptions of a priest and historian Anthony Cogan, 1826–1872* (1993)
Smyth, Bernard T., *The Chinese batch: the Maynooth mission to China: origins, 1911–1920* (1994)
Stacey Sarah Alyn & Véronique Desnain (eds), *Culture and conflict in seventeenth-century France and Ireland* (2004)
Stalley, Roger (ed.), *George Edmund Street and the restoration of Christ Church Cathedral, Dublin* (2000)
Stenson, James B., *Preparing for adolescence: a parents' guide* (1993)
Stewart, Ann M., *Irish art societies and sketching clubs: index of exhibitors, 1870–1980* (1997)

Stewart, Austin, *Coalisland, County Tyrone, in the industrial revolution, 1800–1901* (2002)
Stout, Matthew, *The Irish ringfort* (1997)
Strijbosch, Clara, *The seafaring saint: sources and analogues of the twelfth-century voyage of Saint Brendan* (2000)
Suarez, Federico, *About being a priest* (1979)
Suarez, Federico, *The narrow gate* (1983)
Suarez, Federico, *When the Son of Man comes* (1990)
Sullivan, Mary C. (ed.), *The correspondence of Catherine McAuley, 1818–1841* (2004)
Sullivan, Mary C., *Catherine McAuley and the tradition of mercy* (1995)
Sweeney, Frank, *The murder of Conell Boyle, County Donegal, 1898* (2002)
Sweeney, Michelle, *Magic in medieval romance: from Chrétien de Troyes to Geoffrey Chaucer* (2000)
Sweetinburgh, Sheila, *The role of the hospital in medieval England: gift-giving and the spiritual economy* (2004)
Swift, Roger & Sheridan Gilley (eds), *The Irish in Victorian Britain: the local dimension* (1999)
Talbot, George & Pamela Williams (eds), *Essays in Italian literature and history in honour of Doug Thompson* (2002)
Taylor James W., *The 1st Royal Irish Rifles in the Great War* (2002)
Taylor James W., *The 2nd Royal Irish Rifles in the Great War* (2005)
Taylor, Simon (ed.), *Kings, clerics and chronicles in Scotland, 500–1297: essays in honour of Marjorie Ogilvie Anderson on the occasion of her ninetieth birthday* (2000)
Thompson, F. Glenn, *The uniforms of 1798–1803* (1998)
Trainor, Oliver, *Mother of the Redeemer, Mother of the redeemed* (1988)
Treadwell, Victor, *Buckingham and Ireland, 1616–1628: a study in Anglo-Irish politics* (1998)
Trindade, Ann, *Berengaria: in search of Richard the Lionheart's queen* (1999)
Turpin, John, *Oliver Sheppard, 1865–1941: symbolist sculptor of the Irish cultural revival* (2000)
Twomey, Brendan, *Smithfield and the parish of St Paul, Dublin, 1698–1750* (2005)
Tynan, Michael, *Catechism for Catholics: a guide to living the faith in our time* (1983)
Tynan, Michael, *Catholic instruction in Ireland, 1720–1950* (1985)
Urwin, Margaret, *A Wexford family in the land war; the O'Hanlon Walsh's of Knocktartan* (2002)
Vance, Norman, *Irish literature: a social history: tradition, identity and difference* (1999)
Vandevelde, Karen (ed.), *New voices in Irish criticism 3* (2002)
Walsh, John Edward, *Rakes and ruffians: the underworld of Georgian Dublin* (1979)
Walsh, Micheline Kerney, *An exile of Ireland: Hugh O'Neill, prince of Ulster* (1996)
Walsh, Oonagh (ed.), *Ireland abroad: politics and professions in the nineteenth century* (2003)
Walsh, Paul (ed. and intro. by Nollaig Ó Muraíle), *Irish leaders and learning through the ages* (2003)
Walshe, Eibhear (ed.), *Elizabeth Bowen remembered: the Farrahy addresses* (1998)

Watson, Iarfhlaith, *Broadcasting in Irish: minority, language, radio, television and identity* (2003)
Weber, Paul, *On the road to rebellion: the United Irishmen and Hamburg, 1796–1803* (1997)
Wegemer, Gerald B., *Thomas More: a portrait of courage* (1996)
Weston, Nancy, *Daniel Maclise: an Irish artist in Victorian London* (2000)
Whelan, Bernadette (ed.), *Women and paid work in Ireland, 1500–1930* (2000)
Whelan, Bernadette, *Ireland and the Marshall Plan, 1947–57* (2000)
Whelan, Ruth & Carol Baxter (eds), *Toleration and religious identity: the Edict of Nantes and its implications in France, Britain and Ireland* (2003)
White, Harry, *The progress of music in Ireland* (2005)
Whitehouse, J.C. (ed.), *Catholics on literature* (1997)
Wichert, Sabine (ed.), *From the United Irishmen to twentieth-century unionism: a festschrift for A.T.Q. Stewart* (2004)
Wilson, David A., *United Irishmen, United States: immigrant radicals in the early republic* (1998)
Wood, Kieron, *The High Court: a user's guide* (1998)
Wooding, Jonathan M. (ed.), *The otherworld voyage in Early Irish literature: an anthology of criticism* (2000)
Wright, G.N., *An historical guide to the City of Dublin* (1980)
Zimmermann, Georges Denis, *Songs of Irish rebellion: Irish political street ballads and rebel songs, 1780–1900* (2002)
Zimmermann, Georges Denis, *The Irish storyteller* (2001)

Index